Lecture Notes in Computer Scien

Commenced Publication in 1973
Founding and Former Series Editors:
Gerhard Goos, Juris Hartmanis, and Jan van Leeuwen

María J. Blesa   Christian Blum
Andrea Roli   Michael Sampels (Eds.)

Hybrid
Metaheuristics

Second International Workshop, HM 2005
Barcelona, Spain, August 29-30, 2005
Proceedings

 Springer

Volume Editors

María J. Blesa
Universitat Politècnica de Catalunya
Omega 213 Campus Nord
Jordi Girona 1-3, 08034 Barcelona, Spain
E-mail: mjblesa@lsi.upc.edu

Christian Blum
Universitat Politècnica de Catalunya
Omega 112 Campus Nord
Jordi Girona 1-3, 08034 Barcelona, Spain
E-mail: cblum@lsi.upc.edu

Andrea Roli
Università degli Studi "G. D'Annunzio"
Viale Pindaro 42, 65127 Pescara, Italy
E-mail: a.roli@unich.it

Michael Sampels
Université Libre de Bruxelles
IRIDIA CP 194/6
Avenue Franklin D. Roosevelt 50, 1050 Bruxelles, Belgium
E-mail: msampels@ulb.ac.be

Library of Congress Control Number: 2005930813

CR Subject Classification (1998): F.2, F.1, G.1.6, G.1.2, G.2.1, I.2

ISSN	0302-9743
ISBN-10	3-540-28535-0 Springer Berlin Heidelberg New York
ISBN-13	978-3-540-28535-9 Springer Berlin Heidelberg New York

Springer is a part of Springer Science+Business Media

springeronline.com

© Springer-Verlag Berlin Heidelberg 2005
Printed in Germany

Typesetting: Camera-ready by author, data conversion by Scientific Publishing Services, Chennai, India
Printed on acid-free paper   SPIN: 11546245   06/3142   5 4 3 2 1 0

Preface

Combinatorial optimization and in particular the great variety of fascinating problems that belong to this area have attracted many researchers for more than half a century. Due to the practical relevance of solving hard real-world problems, much research effort has been devoted to the development of heuristic methods aimed at finding good approximate solutions in a reasonable computation time. Some solution paradigms that are not specific for one particular problem have been deeply studied in the past, and the term *metaheuristic* is now common for such optimization heuristics. Several metaheuristics – simulated annealing, genetic and evolutionary algorithms, tabu search, ant colony optimization, scatter search, iterated local search, and greedy randomized adaptive search procedures being some of them – have found their own research communities, and specialized conferences devoted to such techniques have been organized.

Plenty of classical hard problems, such as the quadratic assignment problem, the traveling salesman problem, problems in vehicle routing, scheduling, and timetabling, etc., have been tackled successfully with metaheuristic approaches. Several thereof are currently considered state-of-the-art methods for solving such problems. However, for many years the main focus of research was on the application of single metaheuristics to given problems. A tendency to compare different metaheuristics against each other could be observed, and sometimes this competition led to thinking in stereotypes in the research communities.

In recent years, it has become evident that the concentration on a sole metaheuristic is rather restrictive for advancing the state of the art of tackling both academic and practical optimization problems. A skilled combination of concepts stemming from different metaheuristics can lead to more efficient behavior and greater flexibility. Also the hybridization of metaheuristics with other techniques known from classical artificial intelligence areas, for example data mining, machine learning, etc., can be very fruitful. Further, the incorporation of typical operations research techniques, such as integer or linear programming techniques, branch-and-bound techniques, etc., can be very beneficial. Combinations of metaheuristic components with components from other metaheuristics or optimization strategies from artificial intelligence or operations research are called *hybrid metaheuristics*.

The design and implementation of hybrid metaheuristics raises problems going beyond questions about the composition of a single metaheuristic. A careful analysis of the single components is very important for their interaction. Choice and tuning of parameters is more important for the quality of the algorithms than before. Different concepts of interaction at low-level and at high-level are studied. As a result, the design of experiments and the proper statistical evaluation is more important than before.

The growing interest in research on hybrid metaheuristics and the observed tendency that techniques uncommon in standard metaheuristic research had become of special importance led us to organize a workshop devoted to this particular area. The First International Workshop on Hybrid Metaheuristics (HM 2004) took place in August 2004 in Valencia, Spain. The proceedings of HM 2004 were published as ISBN 3-00-015331-4, and they are available online (http://iridia.ulb.ac.be/~hm2004/proceedings). They contain 13 papers selected from 25 submission.

The success of the first workshop encouraged us to organize a Second International Workshop on Hybrid Metaheuristics (HM 2005) in Barcelona, Spain. The program committee of HM 2005 consisted of 23 researchers and practitioners mostly coming from the hybrid metaheuristics research community, but also from related areas, and from business. We received 37 paper submissions to HM 2005. Each submitted paper was sent to at least three reviewers. We are very grateful to the members of the program committee and the additional reviewers for the effort they made carefully examining the papers and for the many valuable comments and suggestions they gave to the authors. Based on their comments, we finally accepted 13 submissions for publication and for presentation at HM 2005, resulting in an acceptance rate of roughly 35 %. The selection of papers was rather strict in order to guarantee the high quality of the proceedings and the workshop itself. We would like to thank all the authors for their interest in our workshop.

We believe that the combination of elements coming from different metaheuristics, and from methods from both artificial intelligence and operations research, promises to become one of the main tracks of research in applied artificial intelligence. It seems to be a propitious and rewarding alternative to the still existing mutual contempt between the fields of exact methods and approximate techniques, and also to the competition between the different schools of metaheuristics, which sometimes focused more on a proof of concept than on good results.

Still, we have to realize that research on hybrid metaheuristics is mostly based on experimental methods, thus being probably more related to natural sciences than to computer science. It can be stated that both the design and the evaluation of experiments have still not reached the standard they have in physics or chemistry for example. The validity of analyses of experimental work on algorithms is a key aspect in hybrid metaheuristics, and the attention of researchers to this aspect seems to be important for the future of the field.

We observed that the subject matter covered by the submissions to HM 2005 already showed a slight shift from academic to practical and real-world optimization problems. To mention the practical applicability of hybrid metaheuristics as a *raison d'être* seems no longer to be an academic excuse but a real fact. It would be a great success if the growing interdisciplinary cooperation could help to continue the trend of contributing more and more to the area of real-world optimization problems by hybrid metaheuristic approaches.

June 2005

María J. Blesa[1]
Christian Blum[2]
Andrea Roli
Michael Sampels

[1] María J. Blesa acknowledges partial support by the FET Programme of the EU under contract number IST-2004-15964 (AEOLUS) and COST-295 (DYNAMO), and also by the Spanish Science and Technology Ministry (MCyT) under contract number TIC2002-04498-C05-03 (TRACER).
[2] Christian Blum acknowledges the support of a post-doctoral fellowship under the "Juan de la Cierva" program of the Spanish Ministry of Science and Technology.

Organization

Workshop Chairs

María J. Blesa	Universitat Politècnica de Catalunya, Barcelona, Spain
Christian Blum	Universitat Politècnica de Catalunya, Barcelona, Spain
Andrea Roli	Università degli Studi "G. D'Annunzio", Chieti-Pescara, Italy
Michael Sampels	Université Libre de Bruxelles, Belgium

Program Committee

Enrique Alba	Universidad de Málaga, Spain
Mauro Birattari	Université Libre de Bruxelles, Belgium
Thomas Bousonville	ILOG, France
Kirsten Bremke	AT Kearney, Germany
Ralf Bruns	Fachhochschule Hannover, Germany
Óscar Cordón	Universidad de Granada, Spain
Carlos Cotta	Universidad de Málaga, Spain
Luca Di Gaspero	Università degli Studi di Udine, Italy
Marco Dorigo	Université Libre de Bruxelles, Belgium
Filippo Focacci	ILOG, France
Joshua Knowles	University of Manchester, England
Frank Köster	Universität Oldenburg, Germany
Andrea Lodi	Università di Bologna, Italy
Vittorio Maniezzo	Università degli Studi di Bologna, Italy
Monaldo Mastrolilli	IDSIA, Switzerland
Daniel Merkle	Universität Leipzig, Germany
Bernd Meyer	Monash University, Australia
Michela Milano	Università di Bologna, Italy
Olivia Rossi-Doria	Università di Padova, Italy
Andrea Schaerf	Università degli Studi di Udine, Italy
Thomas Stützle	Technische Universität Darmstadt, Germany
El-Ghazali Talbi	École Polytechnique Universitaire de Lille, France
Fatos Xhafa	Universitat Politècnica de Catalunya, Spain

Additional Referees

Prasanna Balaprakash, Max Manfrin, Jose Santamaría, Andrew Tuson

Sponsoring Institutions

Universitat Politècnica de Catalunya, Spain
Turisme de Barcelona (http://www.barcelonaturisme.com)
Direcció General de Turisme de Catalunya (http://www.gencat.net/ctc/turisme)
Transports Metropolitans de Barcelona (http://www.tmb.net)
Caves Freixenet
Caja Madrid
Punto Blanco

Table of Contents

Comparing Parallelization of an ACO: Message Passing vs. Shared Memory

Pierre Delisle[1], Marc Gravel[1], Michaël Krajecki[2], Caroline Gagné[1], and Wilson L. Price[3]

[1] Département d'informatique et de mathématique,
Université du Québec à Chicoutimi, Chicoutimi, Québec, Canada, G7H 2B1
{pierre_delisle, marc_gravel, caroline_gagne}@uqac.ca
[2] Département de Mathématiques et Informatique,
Université de Reims Champagne-Ardenne – F-51687 Reims Cedex 2, France
michael.krajecki@univ-reims.fr
[3] Faculté des Sciences de l'administration,
Université Laval, Québec, Canada, G1K 7P4
wilson.price@fsa.ulaval.ca

Abstract. We present a shared memory approach to the parallelization of the Ant Colony Optimization (ACO) metaheuristic and a performance comparison with an existing message passing implementation. Our aim is to show that the shared memory approach is a competitive strategy for the parallelization of ACO algorithms. The sequential ACO algorithm on which are based both parallelization schemes is first described, followed by the parallelization strategies themselves. Through experiments, we compare speedup and efficiency measures on four TSP problems varying from 318 to 657 cities. We then discuss factors that explain the difference in performance of the two approaches. Further experiments are presented to show the performance of the shared memory implementation when varying numbers of ants are distributed among the available processors. In this last set of experiments, the solution quality obtained is taken into account when analyzing speedup and efficiency measures.

1 Introduction

Many interesting combinatorial optimization problems belong to the class said to be NP-hard [1] and therefore cannot be solved in polynomial time by exact algorithms, be they sequential or parallel. Metaheuristics offer a solution approach which, without guaranteeing optimality, is generally able to produce good solutions. Because these methods often require long computation times and considerable working memory, parallelization would seem to be a promising avenue for performance improvement. The field of parallel metaheuristics is, however, young and the transition between a conventional sequential metaheuristic and an efficient parallel implementation is neither simple nor obvious.

Works describing parallel approaches to combinatorial optimization problems are generally based on one of the two following paradigms: message passing and shared

M.J. Blesa et al. (Eds.): HM 2005, LNCS 3636, pp. 1–11, 2005.

memory. However, most of the recent literature is based on the message passing model which is better known and more mature. The recent emergence of shared memory multiprocessors such as the SMP (Symmetric Multiprocessor) revived interest in the shared memory model, but shared memory implementations of metaheuristics are rare and it is therefore difficult to evaluate the potential of this approach. The aim of this paper is to propose a shared memory approach to the parallelization of the Ant Colony Optimization (ACO) metaheuristic and to compare its performance with a message passing implementation described by Randall & Lewis [2]. The ACO algorithm they use as a basis for their implementation is very similar to ours, as is the algorithmic approach to parallelization. For these reasons, we found a comparison of the experimental results of the two approaches to be particularly interesting and appropriate.

2 The ACO and Literature Review on Its Parallelization

The version of the ACO used by Randall & Lewis is described in Fig. 1. It is the ACO proposed by Dorigo & Gambardella [3] for the solution of the travelling salesman problem. However, Randall & Lewis did not use a candidate list in the transition rule to limit computations and no local search is incorporated in the algorithm. It is important to define the specific version of the algorithm being used to ensure a fair comparison of the two parallel implementations of the ACO.

Briefly, the description of the metaheuristic is as follows: the initial cycle of the algorithm starts with a random choice of the starting city for each of the ants. For each ant, tours are then constructed by adding cities one at a time. A certain number of ants (m), which may be considered to be independent agents, construct tours simultaneously. Each city is added to the tour according to a transition rule (Expression 1) which takes into account the visibility d (distance) as well as the accumulation of the pheromone trail (t) which is a form of memory of the quality of solutions found so far. As a new city is added to the tour, a local update of the pheromone trail (Expression 2) is carried out to reduce the probability of other ants repeating the same choice. When all ants have completed a tour, a global update of the pheromone trail is made (Expression 3) using the best tour (L^+) found in the cycle. If required, the best solution found so far is updated and a new cycle commences.

There are few parallel implementations of the ACO in the literature because it is a relatively new metaheuristic. Using the message passing model, Bullnheimer et al. [4] proposed two parallel versions of the Ant System: a synchronous version and a partially asynchronous version. The parallel synchronous version may be thought of as a low-level parallelization whose aim is to accelerate the algorithm by the generation of parallel ants in a master-slave fashion. At each cycle, the master broadcasts the pheromone trail matrix to the slaves and each slave constructs a tour which, along with its tour-length, is reported to the master. A synchronization of the process is carried out at each iteration. To reduce communication costs in the synchronous version, these authors developed a partially synchronous version where a number of cycles of the sequential algorithm are carried out independently on different processors. Following these "local" iterations, a global update is carried out by the master. Subsequent to a comparison of the performance of these two versions of the algorithm, the authors concluded that the partially asynchronous version is preferable because it allows a major reduction in the frequency and volume of communications. Note that these algorithms were not implemented on a parallel architecture and it is difficult to determine the efficiency of their parallelization scheme.

Set the pheromone trail matrix $\tau(0)$ at $\tau_0 = (n \, L_{nn})^{-1}$ for each pair of city ij;
FOR $t = 0$ to t_{Max} DO
    **** *Ant construction and local update of the pheromone trail* ****
    FOR $k = 1$ to m DO
      Place ant k on a randomly chosen city;
    End FOR
    FOR $i = 1$ to $n-1$ DO
      FOR $k = 1$ to m DO
        Choose the next city $j \in J_i^k$ to visit according to :

$$j = \begin{cases} \underset{\ell \in Tabu_k}{\mathrm{argmax}} \left\{ [\tau_{i\ell}(t)] \cdot \left[\frac{1}{d_{i\ell}} \right]^\beta \right\} & \text{if } q \leq q_0 \\ \\ J & \text{if } q > q_0 \end{cases}$$

        where J is chosen according to the probability: (1)

$$p_{ij}^k(t) = \frac{[\tau_{ij}(t)] \cdot \left[\dfrac{1}{d_{ij}} \right]^\beta}{\displaystyle\sum_{\ell \in Tabou_k} [\tau_{i\ell}(t)] \cdot \left[\dfrac{1}{d_{i\ell}} \right]^\beta}$$

      Local update of the pheromone trail for (i, j):

$$\tau_{ij}(t) = (1 - \rho_\ell) \cdot \tau_{ij}(t) + \rho_\ell \cdot \tau_0 \qquad (2)$$

        Update the length tour L_k with the addition of the city j;
      End FOR
    End FOR
    *** *Best solution and global update of the pheromone trail* ***
    Update of L^+, the best solution so far found;
    Global update of the pheromone trail using L^+:

$$\tau_{ij}(t+1) = (1 - \rho_g) \cdot \tau_{ij}(t) + \rho_g \cdot \Delta\tau_{ij}(t) \qquad (3)$$

End FOR

Fig. 1. Description of the sequential ACO (Randall & Lewis)

Stützle [5] presents two strategies for parallelization using a message passing architecture: the execution of multiple copies of the same algorithm, and the acceleration of a single copy of the algorithm. In the first case, one may use the same search parameters or vary them for each of the parallel runs. While the author did not obtain significant differences between the two approaches when he used them in applying the Max-Min Ant System to the travelling salesman problem, he suggests that performance differences could occur for other problems. The second case, that aimed at the acceleration of a single execution of the algorithm, is similar to the synchronous version described by Bullnheimer *et al.*, but in addition it uses a local search procedure. An efficient parallelization can therefore be obtained by applying a local search in parallel to solutions previously generated.

Talbi *et al.* [6] successfully solved the quadratic assignment problem using a parallel approach. They proposed a parallel model of the ACO similar to the synchronous model of Bullnheimer *et al.*, however they used Tabu Search as a local improvement method.

Michel & Middendorf [7] proposed an island model of the ACO where colonies exchange the best solutions they have found after a fixed number of cycles. When a colony receives a solution that is better than its own best solution, an update is carried out. This new information influences the colony because the pheromone trail updates are carried out using this solution.

Middendorf *et al.* [8] studied four strategies for the exchange of information among multiple ant colonies. They show that it may be advantageous for the colonies to avoid exchanging too much information and to avoid too frequent exchanges. Abandoning the idea of exchanging complete pheromone trail matrices, they based their strategies on the exchange of a single solution at a time. They thus obtained efficient parallel implementations.

Randall & Lewis [2] proposed five strategies for the parallelization of the ACO and give detailed results obtained using one of these strategies. In the following section, we describe this strategy and the results obtained by Randall & Lewis will be used as a basis for comparison with a shared memory parallelization approach that we ourselves propose.

3   The Randall & Lewis *Message Passing* Parallelization of the ACO

Randall & Lewis [2] developed a parallel ACO to solve the travelling salesman problem on a distributed memory architecture. Their approach of the "parallel ant" type is an internal parallelization composed of a master processor and multiple slave processors. Each of the slaves is assigned to an ant and is tasked with the construction of a tour. The master receives the input from the user, assigns a starting city to each ant, carries out the local and global updates of the pheromone trail and produces the output results. Fig. 2 and 3 describe the activities of the master and the slaves in pseudo-code.

It must be noted that the algorithm of Randall & Lewis assigns only one ant to each processor. In their numerical experiments, the number of processors varies from 2 to 8, $\rho_g = 0.1$, $q_0 = 0.9$, $t_{Max}=1000$ and $\beta = 2$.

Broadcast the algorithm parameters to each ant
Broadcast the d matrix to each ant
FOR $t = 0$ to t_{Max} DO
    Place ant k on a randomly chosen city
    Send each initial city to each ant
    FOR $i = 1$ to n-1 DO
        Receive each ant's next city and add it to the corresponding tour
        Update the pheromone matrix τ using the local update rule
        Broadcast m pheromone update of the matrix τ to each ant
    End FOR
    Receive the length tour L from each ant
    Update of L^+, the best solution so far found
    Global update of the pheromone trail using L^+
    Broadcast the pheromone update of the matrix τ to each ant
End FOR
Broadcast *the termination condition* to each ant
Print the shortest tour and its length L^+

Fig. 2. Pseudo-code for the master processor

Receive the algorithm parameters from the master
Receive the matrix d from the master
Initialize the pheromone matrix τ
WHILE *the termination condition* is met DO
   Initial_City = City = Initial city receive from the master
   FOR $i = 1$ to n-1
      Next_City = Choose the next city according the equation (1)
      Send *Next_City* to the master
      $L = L + d\ (City, Next\_City)$
      $City = Next\_City$
      Receive the m pheromone update of the matrix τ from the master
   End FOR
   $L = L + d\ (Next\_City, Initial\_City)$
   Send L to the master
   Receive the pheromone update of the matrix τ from the master
   Receive the *termination information signal* from the master
End WHILE

Fig. 3. Pseudo-code for the slave processors

4 A *Shared Memory* Parallelization of the ACO

In this section, we propose a shared memory parallelization approach of the internal, fine-grain type. Certain inherent constraints on the sequence of operations must be respected in the parallel version so as to preserve the integrity of the search for solutions. The distribution of ants to the processors must allow the simultaneous construction of solutions and the local update must be carried out each time a city is added to a tour. Using a local update strictly equivalent to that used in the sequential algorithm would require an onerous synchronization and would be a severe limit on the potential of the parallelization. We have examined the quality of solutions to problems having as many as 2,000 cities [10] and found that the search process is not adversely affected. We therefore conclude that it is desirable to relax this requirement. Moreover, certain precedence relations for the update of the best solution so far found must be preserved. This update can only be done one processor at a time if one is to avoid data integrity problems. Finally, one must ensure that all the processors have constructed, evaluated and compared their solutions to the best known solution before a global update is undertaken within a given cycle.

The parallelization proposed is described in Fig. 4. It distinguishes three groups of operations that are treated separately in a single parallel region. The first group of

```
Initialize τ
FOR t = 0 to t_Max DO
    Choose randomly the first city for each of the m ants
    *** Start a parallel region with m processors ***
        NoProc = Index of the processor
        FOR i = 1 to n  DO
                FOR k = 1 to m  DO
                        If NoProc = = k modulo m
                            Choose the next city to visit for ant k
                            Update the length tour L_k with the city added to the tour
                            *** Critical Zone ***
                            Local update of τ by the processor k
                End FOR
        End FOR
        FOR k = 1 to m DO
                If NoProc = = k modulo m
                    If L_k < L⁺
                    *** Critical Zone ***
                    Update of L⁺, the best solution so far found
        End FOR
            *** Synchronization barrier ***
        FOR i = 1 to n  DO
                If NoProc = = i modulo m
                FOR j = 1 to n DO
                    Apply equation (3) for edge (i, j)
                End FOR
        End FOR
        Global update of the pheromone trail using L⁺
        *** End the parallel region ***
    End FOR
    Print L⁺
```

Fig. 4. Pseudo-code for the parallelization in shared memory

parallel operations equitably distributes the ants to processors and imposes a desynchronization of the local update of the pheromone trail. The second group of parallel operations carries out the update of the best solution known so far. By again equitably distributing the ants among processors, we allow a processor that has completed constructing its solutions to update the best known solution even if the other processors have not yet completed the first group of operations. The processor in question compares its best solution to the best solution stored in shared memory and updates it if required. This update must, on the other hand, be carried out within the critical zone to guarantee that a single processor will write to this data structure at a given moment. Finally the third group of parallel operations carries out the global update of the pheromone trail by uniformly distributing the update of the lines of the matrix to different processors. However, this procedure must be preceded by a synchronization barrier that ensures that all the processors have finished the treatment of all the ants and that the best solution has, in fact, been updated. Let us note that, in the three groups of operations in the parallelization, the computing load associated with the tasks that are carried out in parallel is regular and that a static scheduling is therefore sufficient.

This parallelization in shared memory may be used with any number of ants. If we use a number of ants equal to the number of processors, we obtain the configuration used by Randall & Lewis. The numerical experiments were carried out using the same conditions as used in the message passing approach so as to obtain a valid comparison.

5 Experimentation and Results

The experiments of Randall & Lewis were carried out on eight travelling salesman problems having between 24 and 657 cities, using an IBM SP2 computer having 18 266Mhz RS6000 model 590 processors. However, in the current comparison, only the four biggest problems were retained. The experiments in shared memory were carried out using 16 375Mhz Power3 processors on a NH2 node of an IBM/P 1600. We seek an approach that is as general as possible and not particular to a given type of computer, and our implementation of the parallelization strategy was facilitated through the use of OpenMP which is a new tool that allows the development of programs portable to a number of shared memory parallel computers.

Table 1 compares speedup and efficiency, which are the two performance measures used by Randall & Lewis in their paper, for both approaches. However, it is important to note that our performance measures were not produced by following the exact same guidelines. For each problem, Randall & Lewis performed a single execution with a fixed seed and calculated speedup with the following formula:

$$\text{Speedup} = \frac{\text{Time to solve a problem with the fastest serial code on a specific parallel computer}}{\text{Time to solve the problem with the parallel code using } p \text{ processors on the same computer}}$$

The numerator was measured by CPU time and the denominator was measured by wall clock time. We used the same formula and time measures to calculate speedup. However, our results are the average of ten trials with different seeds, which seemed to be a realistic setup for stochastic algorithms such as ACO. The method of Randall & Lewis for calculating speedup follows more strictly the guidelines outlined by Barr and Hickman [9], however our method is also generally accepted. For this reason, the results presented here should be interpreted as showing trends rather than comparing strict numerical measures.

Table 1. Speedup and efficiency

Problems	Message-passing (Randall & Lewis)							Shared-memory						
	Number of processors							Number of processors						
	2	3	4	5	6	7	8	2	3	4	5	6	7	8
lin318	1.20	1.44	1.44	1.59	1.61	1.53	1.58	1.65	2.39	3.09	3.64	4.15	4.63	4.77
(318 cities)	0.60	0.48	0.36	0.32	0.27	0.22	0.20	0.83	0.80	0.77	0.73	0.69	0.66	0.60
pcb442	1.42	1.62	1.93	2.18	2.31	2.31	2.35	1.71	2.47	3.26	4.02	4.57	5.24	5.55
(442 cities)	0.71	0.54	0.48	0.44	0.38	0.33	0.29	0.86	0.82	0.81	0.80	0.76	0.75	0.69
rat575	1.56	1.78	2.10	2.55	2.77	3.02	3.08	1.78	2.54	3.23	3.95	4.62	5.14	5.74
(575 cities)	0.78	0.59	0.52	0.51	0.46	0.43	0.38	0.89	0.85	0.81	0.79	0.77	0.73	0.72
d657	1.67	1.95	2.32	2.89	3.25	3.29	3.30	1.74	2.62	3.39	4.12	4.83	5.36	6.14
(657 cities)	0.83	0.65	0.58	0.58	0.54	0.47	0.41	0.87	0.87	0.85	0.82	0.80	0.77	0.77

We note that the performance of the shared memory approach is superior in all cases. The measures of speedup are larger and increase more rapidly when the number of processors is increased. For efficiency, the measures are also better and their decrease is slower. A number of factors can explain these differences in performance :

- In the algorithm of Randall & Lewis, for each local update, all of the slaves send a message to the master and wait for the master's broadcast of the updates to carry out before continuing the construction of their respective solutions. This processor inactivity is minimized in the shared memory approach by the desynchronization of the local update;
- Management of the access to shared memory and of the synchronizations may be more efficient at the software and hardware levels than is the case for message passing routines for this type of parallelization;
- The parallel computer used in the shared memory experiments is more technologically advanced and is probably better at managing parallel computations;
- Various other technical factors such as the compiler and the quality of the code may influence performance.

Even if it is not possible to draw certain conclusions concerning the actual relevance of each of these factors, it would seem reasonable to believe that because of some combination of them, the shared memory approach performs better.

In Delisle et al. [10], it was established that it is generally preferable to increase the workload on the processors by assigning several ants to each. It is therefore interesting to consider numerical experiments that do not use a single ant per processor as did Randall & Lewis. The 657 cities problem was therefore solved using 10, 20 and 40 ants shared equally among the processors. The number of cycles carried out in each experiment was, however, reduced so as to maintain the total number of operations obtained in the previous version. Table 2 presents a comparison of the results obtained. We note that the increase in the number of ants assigned to each processor results in a significant increase in the parallel performance. When $m = 40$, the measures of efficiency vary from 82% to 92% regardless of the number of processors used.

Table 2. Solution quality (% of the best known solution), speedup and efficiency obtained by varying the number of ants per processor for the shared memory approach (Problem d657)

m	Number of processors						
	2	3	4	5	6	7	8
Equal to the number of processors	0.294	0.289	0.280	0.292	0.294	0.303	0.303
	1.74	2.62	3.39	4.12	4.83	5.36	6.14
	0.87	0.87	0.85	0.82	0.8	0.77	0.77
10	0.342	0.344	0.343	0.333	0.348	0.324	0.321
	1.82	2.34	2.99	4.09	4.13	4.18	4.29
	0.91	0.78	0.75	0.82	0.69	0.60	0.54
20	0.368	0.357	0.340	0.355	0.341	0.334	0.332
	1.87	2.61	3.58	4.29	4.49	5.63	5.65
	0.94	0.87	0.90	0.86	0.75	0.80	0.71
40	0.371	0.362	0.352	0.344	0.356	0.348	0.348
	1.84	2.66	3.56	4.35	4.97	5.74	6.71
	0.92	0.89	0.89	0.87	0.83	0.82	0.84

Table 3. Solution quality (% of the best known solution), speedup and efficiency obtained by varying the number of ants per processor for the shared memory approach with a constant number of cycles (Problem d657)

m	Number of processors						
	2	3	4	5	6	7	8
10	0.316	0.311	0.300	0.319	0.304	0.327	0.320
	1.84	2.34	3.04	4.14	4.18	4.21	4.31
	0.92	0.78	0.76	0.83	0.70	0.60	0.54
20	0.320	0.321	0.325	0.309	0.320	0.331	0.318
	1.87	2.65	3.63	4.49	4.65	5.78	5.79
	0.93	0.88	0.91	0.90	0.77	0.83	0.72
40	0.322	0.324	0.329	0.321	0.315	0.320	0.313
	1.88	2.69	3.68	4.56	5.17	5.96	7.07
	0.94	0.90	0.92	0.91	0.86	0.85	0.88

As to the solution quality, the reduction in the number of cycles to preserve the same number of operations causes a slight reduction in the average quality of solutions obtained. Quality of a particular solution is measured as the percentage gap between the value of the best known solution for this problem (48912) and the value of the solution considered. If the number of ants is fixed at 10, to preserve the same number of evaluations, the number of cycles goes from 200 (for 2 processors) to 800 (for 8 processors). If the number of ants is fixed at 20, for the same reason, the number of cycles will go from 100 (for 2 processors) to 400 (for 8 processors). For 40 ants, the range is from 50 cycles (for 2 processors) to 200 cycles (for 8 processors). If the number of cycles of the algorithm remained constant at 1000 whatever the number of processors used, the results of Table 3 allow us to see that the measures of efficiency increase again and that the quality of the solutions increases as well. The configuration where $m=10$ allows us to obtain solutions close to the quality of those obtained in Table 2 where the number of ants is equal to the number of processors.

In any case numerical experiments where the parameter values are varied allow us to draw better conclusions concerning quality. In Delisle *et al.* [10], it was also shown that the use of local search methods improved solution quality and preserved good values of efficiency and of speedup.

6 Conclusion

The objective of the comparisons presented in this work is not to determine the superiority of one parallel implementation over another. As pointed out in this article, many technical factors can influence performance and prevent the drawing of definitive conclusions about the quality of the parallelization approaches discussed. By reproducing as closely as possible the parallelization context of an ACO via message passing, we sought to show that the shared memory architecture offers a competitive avenue for the parallelization of this metaheuristic. We also wanted to show that it is possible to obtain good results in such an environment while respecting the constraints of the sequential approach. We also showed that a more realistic choice of parameter values as to the number of ants used by the algorithm has a positive impact on the performance when the number of processors varies from 2 to 8. This is a configuration found on currently available and reasonably priced shared memory parallel computers.

The comparisons in this paper use a form of internal parallelization, but another popular parallelization strategy for the ACO is the establishment of multiple colonies which may be cooperative or independent. This approach has already been studied in the message passing context [5][8]. In future work, we will explore a parallelization using multiple colonies in a shared memory environment and we will compare results with existing approaches. A further approach that we suggest is the creation of hybrid *shared memory/message passing* algorithm. This approach would allow, for example, the evolution of a number of cooperative colonies that communicate by message passing on the nodes of an SMP cluster. Each colony could therefore be internally parallelized on each multiprocessor, shared memory node. The potential of such architectures could allow the solution of large scale problems with reasonable computing times.

References

1. Garey, M.S., Johnson, D.S.: Computer and Intractability : A Guide to the Theory of NP-Completeness. New York, W.H. Freeman and Co. (1979)
2. Randall, M., Lewis, A.: A Parallel Implementation of Ant Colony Optimization. Journal of Parallel and Distributed Computing, Academic Press Inc, 62, 2 (2002) 1421-1432

3. Dorigo, M., Gambardella, L.M.: Ant colonies for the Traveling Salesman Problem. BioSystems, 43 (1997) 73-81
4. Bullnheimer, B., Kotsis, G., Strauss, C.: Parallelization Strategies for the Ant System. In: R. De Leone, A. Murli, P. Pardalos, and G. Toraldo (Ed.), High Performance Algorithms and Software in Non-linear Optimization , Kluwer Academic Publishers (1998) 87-100
5. Stützle, T.: Parallelization Strategies for Ant Colony Optimization. In: Proceedings of Parallel Problem Solving from Nature -- PPSN-V, Amsterdam, Lecture Notes in Computer Sciences, Springler Verlag, A.E. Eiben, T. Bäck, M. Schoenauer, and H.P Schwefel (ed.) (1998) 722-731
6. Talbi, E.-G., Roux, O., Fonlupt, C., Robillard, D.: Parallel Ant Colonies for Combinatorial Optimization Problems. In: BioSP3 Workshop on Biologically Inspired Solutions to Parallel Processing Systems, IEEE IPPS/SPDP'99 (Int. Parallel Processing Symposium / Symposium on Parallel and Distributed Processing), J. Rolim, San Juan, Puerto Rico, USA, Springer-Verlag (1999)
7. Michel, R., Middendorf, M.: An Ant System for the Shortest Common Supersequence Problem. In New Ideas in optimization, D. Corne, M. Dorigo, F. Glover (ed.) (1999) 51-61
8. Middendorf, M., Reischle, F., Schmeck, H.: Information Exchange in Multi Colony Ant Algorithms. In Parallel and Distributed Computing, Proceedings of the 15 IPDPS 2000 Workshops, J.Rolim, G. Chiola, G. Conte, L.V. Mancini, O.H. Ibarra and H. Nakano (Ed.), Cancun, Mexico, Lecture Notes in Computer Sciences, Springer-Verlag (2000) 645-652
9. Barr, H., Hickman, B.: Reporting computational experiments with parallel algorithms: Issues, measures and experts' opinions. ORSA Journal of Computing, 5 (1993) 2-18.
10. Delisle, P., Gravel, M., Krajecki, M., Gagné, C., Price, W.L.: A Shared Memory Parallel Implementation of Ant Colony Optimization. Working Paper, Université du Québec à Chicoutimi (2005)

An LP-Based Hybrid Heuristic Procedure for the Generalized Assignment Problem with Special Ordered Sets

Alan P. French and John M. Wilson

Business School, Loughborough University,
Loughborough LE113TU, England
{a.p.french, j.m.wilson}@lboro.ac.uk

Abstract. The generalized assignment problem with special ordered sets (GAPS2), is the problem of allocating n tasks to m time-periods, where each task must be assigned to a time-period, or shared between two consecutive time-periods. For large values of m and n the NP-hard combinatorial problem GAPS2 becomes intractable for standard mathematical programming software such as CPLEX or XPRESS^{MP} and it is unlikely that a proven optimal solution can be obtained. There is thus a need for heuristic algorithms to solve such problems. It will be shown how a combination of linear programming techniques and two heuristics forming a hybrid can be used to solve GAPS2. Good results, in terms of speed and accuracy, on large problem instances have been obtained. In particular when compared to an existing heuristic for GAPS2, the results look particularly promising.

Keywords: Assignment, generalized assignment, heuristics, hybrid metaheuristic, special ordered sets.

The generalised assignment problem with special ordered sets (GAPS2) is the problem of allocating tasks to time-periods, such that total cost is minimized, where each task must be allocated to one time-period (or split between two consecutive time-periods), more than one task or parts of tasks may be allocated to a particular time-period, but each time period is limited. The problem may be formulated as follows using the notation x_{ij} = proportion of task j allocated to time-period i, c_{ij} = cost of allocating task j to time-period i ($c_{ij} > 0$), b = length of each time-period, a_j = time required to execute task j ($b \geq a_j > 0$). It is assumed that any task can be executed within a time-period, but may be split between two consecutive periods. Tasks are also assumed to be non-preemptable.

GAPS2

$$\text{minimize} \quad z = \sum_{i=1}^{m} \sum_{j=1}^{n} c_{ij} x_{ij} \tag{1a}$$

M.J. Blesa et al. (Eds.): HM 2005, LNCS 3636, pp. 12–20, 2005.

subject to $\displaystyle\sum_{j=1}^{n} a_j x_{ij} \leq b$, $i \in M = \{1,2, \dots ,m\}$ (2a)

$\displaystyle\sum_{i=1}^{m} x_{ij} = 1$, $j \in N = \{1,2,\dots, n\}$ (3)

where $\{\, x_{ij} : i\varepsilon \in M \,\}$ is a special ordered set of type 2 (S2) for each $j \in N$ that is $x_{ij} > 0 \Rightarrow$

$\{(\, x_{i-1,j} = 0$ and $x_{i+1,j} = 1 - x_{ij}\,)$ OR $(\, x_{i+1,j} = 0$ and $x_{i-1,j} = 1 - x_{ij}\,)\}$ (4)

Special ordered sets of type 2 (S2 sets) were introduced by Beale and Tomlin (1970) and further developed by Beale and Forrest (1976). Integer programming (IP) problems containing special ordered sets of variables are solvable using an extension to the branch and bound algorithm and this is provided in standard IP solvers.

In the formulation (1a) gives the expression of total cost which is to be minimized. Constraints (2a) ensure that in each time period the total time used by allocated tasks is within the limit. Constraints (3) – (4) ensure that each task is allocated to a time period, or shared between consecutive ones.

For the purposes of more thorough computational testing the above formulation will be generalized slightly so that coefficient a_j will be allowed to vary with time-period i (w_{ij} will be used to denote this) and not all time-periods will be of the same length (b_i will be used to denote this, where $w_{ij} \leq b_i$, $\forall j$). The objective function will also be generalized so that it can refer to profit (p_{ij}) as well as cost. Thus (1a) and (2a) are replaced by:

maximize (or minimize) $z = \displaystyle\sum_{i=1}^{m} \sum_{j=1}^{n} p_{ij} x_{ij}$ (1)

subject to $\displaystyle\sum_{j=1}^{n} w_{ij} x_{ij} \leq b_i$, $i \in M = \{1,2, \dots ,m\}$ (2)

GAPS2 was introduced by de Farias Jr et al.(2000) and is an extension of the NP-hard generalized assignment problem (GAP). A formulation of GAP would be (1)-(3) with (4) replaced by the requirement $x_{ij} = 0$ or 1 $i \in M$, $j \in N$. Much research has been conducted on GAP. Wilson (2002) reviews many earlier papers so we shall only mention here the most recent ones. Tabu search approaches to GAP have been developed by Diaz and Fernandez (2001) and Higgins (2001), who tackles large problems, a path re-linking method has been developed by Yagiura et al. (2002) and an ejection chain approach by the same authors in Yagiura et al. (2004a), a paper by Nauss (2002) develops heuristic approaches but also embeds them in an exact approach, and finally a paper by Haddadi and Ouzia (2004) uses a Lagrangian approach. These papers are indicative that research into both exact and heuristic methods for GAP is still active. Further related work on an extension of GAP appears in Yagiura et al (2004b).

GAPS2 has been less extensively studied. Wilson (2002) presents a straightforward heuristic for large instances of the problem, in contrast to de Farias Jr et al. (2000) who solved smaller instances of the problem by an exact method based on a polyhedral approach. Applications of the problem, as discussed in de Farias Jr et al. (2000) arose in the context of fibre optics cable production scheduling. A series of jobs had to be scheduled. In that paper sharing of jobs occurred across adjacent time periods. Similar problems will occur if jobs are shared across adjacent machines or tasks between groups of personnel that can be regarded as adjacent in some sense. It is unlikely that the methods developed in de Farias Jr et al. (2000) could be used to solve problem instances of the size intended in this paper.

For reasonably large values of m and n the NP-hard combinatorial problem GAPS2 becomes intractable for standard IP software systems such as XPRESS^{MP}, hence there is a need for the development of heuristic algorithms to solve such problems.

In this paper a new heuristic approach for GAPS2 will be developed. The heuristic is based on ideas developed by Trick (1992) whose heuristic for GAP is acknowledged to be very effective (see for instance Savelsbergh, 1997) provided it can find a feasible solution to a problem instance. The GAPS2 heuristic approach subsequently derived is fast and accurate and considerably outperforms a single-pass heuristic of Wilson (2002). Thus although the ideas behind the heuristic are relatively straightforward, one contribution of this paper is to demonstrate that by adaptation a heuristic that could be very effective in some instances of GAP, but not in others, can be made to be particularly effective for GAPS2. The hybrid heuristic approach developed combines the use of linear programming techniques to guide the search followed by the use of two heuristics to repair and advance solutions.

1 Adapting an Existing GAP Heuristic

The heuristic of Trick (1992) for GAP takes advantage of the fact that the solution to the linear relaxation of GAP (i.e. (1)-(3) with (4) removed) is such that at most m variables will have non-integer values (and this is encouraging because m is normally small compared to n). The heuristic for GAP for a maximize problem may be summarised as follows.

Step 0   Set $x_{ij}^{''} = 0, \quad i \in M; \quad j \varepsilon N$. Let $N^{'} = N$ and $U = \phi$.

Step 1   Solve the linear relaxation of GAP. Let the solution be $\{ x_{ij}^{'} \}$.

Step 2   For each $j \in N^{'}$ if $x_{ij}^{'} = 1 \quad i \in M$ set $x_{ij}^{''} = 1$ and let

$$N^{'} = N^{'} \setminus \{ j \}. \text{ Calculate } s_i = b_i - \sum_{j=1}^{n} w_{ij} x_{ij}^{''}. \text{ If } N^{'} = \phi \text{ stop,}$$

$\{ x_{ij}^{''} \}$ is the solution.

Step 3   For any pair $\{i,j\}$, $j \in N'$, such that $w_{ij} > s_i$, let
$$U = U \cup \{i, j\}.$$

Step 4   Solve the problem **R**

Maximize $\displaystyle\sum_{i=1}^{m} \sum_{j \in N'} p_{ij} x_{ij}$

Subject to $\displaystyle\sum_{j \in N'} w_{ij} x_{ij} \le s_i$, $i \in M$

$\displaystyle\sum_{i=1}^{m} x_{ij} = 1$, $j \in N'$

$x_{ij} = 0$ $\{i, j\} \in U$

and let the solution be $\{ x_{ij}' \}$.

If this problem is infeasible then GAP is declared infeasible.
Go to Step 2.

In order to adapt the above heuristic to solve GAPS2, a change will be required in
Step 3. Step 3 is the method by which fractional values of variables are avoided in
successive iterations, however this step is not usable for GAPS2 without modification
because fractional values of variables are acceptable (provided exactly two adjacent
periods are involved). Step 3 is the key step in ensuring that the problem to be solved
at Step 4 is a new problem by fixing *useless variables* (Trick's terminology) at zero,
otherwise the LP solver would always provide the same x_{ij}' values.

For GAPS2 Step 3 can be generalized to the condition:

If $0 < x_{ij}' < 1$ for some period i and job j and $x_{i-1,j}' = 0$ *and* $x_{i+1,j}' = 0$ then if

$\dfrac{s_k}{w_{kj}} + \dfrac{s_i}{w_{ij}} < 1$ for both $k = i+1$ and $k = i-1$ hold x_{ij} is a useless variable and its

subscripts can be added to the set U. However, there might be no value of i for which
this condition holds even though fractional values of variables remain in problem R.
Therefore the GAP heuristic needs much more extensive adaptation before it can be
used on GAPS2. Before this is considered some further remarks are necessary.

The difficulty of obtaining feasible solutions using Trick's heuristic is alluded to
by Cattryse et al. (1998), but the phenomenon did not occur in the work of that paper
or in Trick (1992) as severely as experienced by the authors of the current paper. This
outcome was slightly unexpected. GAPS2 should be 'easier' to solve than GAP
because certain solutions feasible to GAPS2 would not be feasible for GAP, but it
transpired that our test problems were rendered both GAP-infeasible and GAPS2-
infeasible after the first pass through the GAP heuristic. Thus a rather more extensive
modification of the original heuristic will be required to solve realistic sized GAPS2
instances.

2 A Heuristic for GAPS2

In this section we develop an LP-based heuristic to solve GAPS2. In essence the heuristic proceeds by solving a series of linear relaxations of the GAPS2 (i.e. (1)-(3) with (4) removed). After solving each relaxation, there are three possibilities:

1. A feasible solution to GAPS2 has been found, i.e. any fractional x'_{ij} values satisfy the special ordered sets constraints (4). In this case, the heuristic terminates and passes control to a perturbation phase (described later).
2. The linear relaxation is infeasible. In such cases, the heuristic has failed to find a feasible solution to the related GAPS2 problem and terminates. In practice, this situation was not observed.
3. The linear relaxation is feasible, but there are at least two fractional x'_{ij} values that do not satisfy the special ordered sets constraints (4). In this case, the heuristic tightens the linear relaxation by fixing a number of x'_{ij} values to zero, and solves the tightened linear relaxation.

2.1 Fixing Criteria to Facilitate Heuristic Progress

The heuristic seeks to find a solution by solving a succession of tightened linear relaxations. In order for the heuristic to progress, it needs to fix at least one illegally assigned task-period allocation to zero. A degradation value is calculated for each illegal task-period combination:

$$\deg_{ij} = \frac{p_{ij} b_i x'_{ij}}{w_{ij}}$$

Clearly, it is desirable to minimise the degradation in the objective whilst maximising the amount of space freed up. This is achieved by multiplying the objective degradation by the reciprocal of the slack proportion. x'_{ij} is included in the degradation to take account of the magnitude of change. For example, given two task-period allocations with the same degradation measure, with respective fractional values of 0.1 and 0.9, it is preferable to fix the first at zero.

For each currently illegally assigned task, the task-period allocation with the smallest degradation value is fixed at zero.

2.2 Fixing Criteria to Improve Heuristic Performance

Simply fixing a single illegal task-period allocation at zero proves to be impractical as the problem size grows. Therefore, in order to improve the performance of the heuristic, it is also necessary to fix other task-period combinations to zero.

For each task, all x'_{ij} values are fixed at zero, except for the current fractional (i.e. non-zero) x'_{ij} values and the x'_{ij} values currently equal to one, and the periods immediately preceding and following, thus allowing tasks to be split between two consecutive periods.

2.3 The LP-Based Heuristic

Step 0   Solve GAPS2 (1), (2) and (3) ignoring the S2 constraints (4). Let $\{x_{ij}'\}$ be the solution values.

Step 1   If all tasks are legally assigned then go to Step 6. Else if the LP solution is infeasible, then terminate and report no feasible solution as being found.

Step 2   Let $J = \{j \in N \mid \exists i \in M : x_{ij}' = 1\} \cup \{j \in N \mid \exists i \in M \setminus \{m\} : x_{ij}' + x_{i+1,j}' = 1\}$

Let $J' = N \setminus J$ (i.e. the set of illegally assigned tasks)

Step 3   For each $j \in J'$

Let $M' = \{i \in M \mid x_{ij}' > 0\}$

For each $i \in M'$, let $\deg_{ij} = p_{ij} b_i x_{ij}' / w_{ij}$

Let $i' \in M'$ be the period with the smallest value of $\deg_{ij}$

Let $U' = \bigcup_{i \in M'} \{[i-1, i+1]\}$

Let $F' = (M \setminus U') \cup \{i'\}$

For each $i \in F'$, fix x_{ij} to zero in the LP formulation (1) – (3).

Step 4   For each $j \in J$

Let $M' = \{i \in M \mid x_{ij}' > 0\}$

Let $U' = \bigcup_{i \in M'} \{[i-1, i+1]\}$

Let $F' = (M \setminus U')$

For each $i \in F'$, fix x_{ij} to zero in the LP formulation (1) – (3).

Step 5   Solve the resulting LP formulation and repeat from Step 1.

Step 6   Apply the solution improvement heuristic described in section 3 below.

2.4 The Solution Improvement Heuristic

Upon completion of the LP-based heuristic, a second heuristic is applied. The second heuristic searches for periods with slack, and seeks to find tasks completely allocated to immediately neighbouring periods that could be beneficially split by allocating all or part of the task to the machine with slack capacity. If more than one such period-task combination is found, then the combination leading to the greatest increase in the objective is chosen. The process is repeated until no further improvement is possible.

3   Computational Results

3.1   Initial Testing

The heuristic was now tested on two types of problem instances. Both sets are discussed in Wilson (2002). The instances are available at http://www-staff.lboro.ac.uk/~msjmw/

Good results, in terms of speed and quality, are obtained. For the smallest problems solutions were close to optimal and obtained rapidly. In contrast, using XPRESS^{MP} to obtain the optimal solution requires on average 1000's of branch and bound nodes. For the larger problems, where the optimal solution is unknown, solutions are obtained rapidly in at worst 5.3 CPU seconds and are close to an upper bound value. When the heuristic solution results in a ratio measure of 1.000 when compared to an upper bound that is not guaranteed to be optimal, it is likely that the solution found is optimal.

A second set of problems with correlation between p_{ij} and w_{ij} was generated. These problems are in the style of the Type E generator described in Laguna et al. (1995). The generator is modified slightly because the objective function (1) is maximise. Such problems are designed to provide a harder test for the heuristic because of the correlation effects.

The second set of problems turned out not to be any harder for the heuristic than the first set. Again solutions were obtained rapidly and were near-optimal. Both sets of results significantly improved on those in Wilson (2002), which used the same problem instances, where run times were longer and optimality levels poorer.

3.2   Further Testing

In order to test the algorithm further some standard GAP instances were used to give some comparability. Three sets of 15 problems discussed in Diaz and Fernandez (2001) and Yagiura et al. (2004a) were used. The larger instances of these problems are known to be hard to solve as GAPs. These problems were solved as if they were (minimization) GAPS2 problems. For some of these problems the optimal solution is known and for all of them an upper and lower bound are known (see Wilson, 2002). Table 1 shows the results of these tests on one of the sets. As can be seen from the last column of the table the heuristic is successful in obtaining solutions that are close to optimal. Solutions are, on average, within 0.2% of optimality. It is also apparent that for each problem instance, the differences between the best known GAPS2 solution, the value of the LP relaxation, and the best known value of the GAP, are small. Overall it is promising that the heuristic is able to obtain good quality solutions to GAPS2 rapidly, within 1.5 minutes for even the hardest of the larger problem instances, outperforming a commercial branch and bound code which is less suited to this type of combinatorial problem.

Table 1. Yagiura Type E Problems (minimization)

Problem	DFYIG	XPRESS^MP				Heuristic		ratio
		IP	Bound	Secs	Nodes	Sol	Secs	Sol/IP
E05100	12681	12642.73	12642.73	0	27	12646.63	0.3	1.000
E05200	24930	24924.82	24924.82	0	68	24927.64	0.5	1.000
E10100	11577	11551.09	11551.09	42	16013	11573.15	0.4	1.002
E10200	23307	23295.85	23295.85	25	5467	23310.26	0.6	1.001
E10400	45746	45740.76	45740.76	443	43677	45748.52	1.1	1.000
E15900	102422	102423.1*	102416.79	1679		102429.90	4.0	1.000
E20100	8436	8411.95*	8363.32	1072		8532.86	0.8	1.020
E20200	22379	22376.66*	22358.2	1413		22395.84	1.3	1.002
E20400	44882	44884.13*	44862.21	1190		44912.78	2.3	1.001
E201600	180647	180658.6*	180640.3	1226		180646.40	10.7	1.000
E30900	100438	100477.65*	100413.5	1460		100468.30	10.3	1.001
E40400	44579	44756.27*	44523.71	1536		44634.09	6.6	1.002
E401600	178311	178537.5*	178282.6	1350		178315.30	28.3	1.000
E60900	100177	100995.6*	100103.3	1287		100338.20	28.9	1.002
E801600	176856	**	176781	1086		177006.80	91.84	1.001
							mean	1.002

DFYIG = best heuristic solution to GAP from Diaz and Fernandez (2001) or Yagiura, Ibaraki and Glover (2004a).

- ● = non-optimal IP solution, secs refer to time taken to reach it.

- ● ** = no IP solution found, bound given is LP relaxation. For this instance denominator of Sol/IP replaced by Bound.

4 Conclusions

By adapting a heuristic for GAP to the needs of GAPS2 this paper has shown how it is possible to solve fairly large instances of GAPS2 rapidly and with a consistently high degree of solution quality. When the heuristic was applied to sets of relatively difficult problems from the literature, near-optimal solutions were obtained rapidly with in most cases solutions being within 0.2% of a lower bound.

The GAPS2 heuristic developed can in turn be adapted to provide a method to solve GAP where the problem instances are tightly constrained and the original heuristic of Trick (1992) would not be satisfactory.

References

1. Beale, E.M.L. and Forrest, J.J.H. Global optimization using special ordered sets. Math. Programming 10 (1976) 52-69.
2. Beale, E.M.L. and Tomlin, J.A. Special facilities in a general mathematical programming system for non-convex problems using ordered sets of variables. In: OR 69: Proc. 5th Int. Conf. Oper. Res., Venice 1969 (edited by J. Lawrence) Tavistock Publications, London (1970) pp 447-454.
3. Cattrysse, S., Degraeve, Z. and Tistaert, J. Solving the generalized assignment problem using polyhedral results. Eur. J. Oper. Res. 108 (1998) 618-628.
4. de Farias Jr, I.R., Johnson, E.L. and Nemhauser, G.L. A generalized assigment problem with special odered sets : a polyhedral approach. Math. Programming 89 (2000) 187-203.
5. Diaz, J.A. and Fernandez, E. A tabu search heuristic for the generalized assignment problem. Eur. J. Oper. Res. 132 (2001) 22-38.
6. Haddadi, S. and Ouzia, H. Effective algorithm for the generalized assignment problem. Eur. J. Oper. Res. 153 (2004) 184-190.
7. Higgins, A.J. A dynamic tabu search for large-scale generalised assignment problems. Comput. Oper. Res. 28 (2001) 1039-1048.
8. Laguna, M., Kelly, J., Gonzalez-Velarde, J. and Glover, F. Tabu search for the multilevel generalized assignment problem. Eur. J. Oper. Res. 82, (1995) 176-189.
9. Nauss, R.M. Solving the generalized assignment problem: An optimizing and heuristic approach. INFORMS J. Comp. 15 (2003) 249-266.
10. Savelsbergh, M. A branch-and-price algorithm for the generalized assignment problem. Oper. Res. 45 (1997) 831-841.
11. Trick, M.A. A linear relaxation heuristic for the generalised assignment problem. Naval Res. Logist. 39 (1992) 137-151.
12. Wilson, J.M. An algorithm for the generalized assignment problem with special ordered sets. Accepted by J. Heuristics. (2002)
13. XPRESS[MP], Dash Optimization, Blisworth, Northamptonshire, England.
14. Yagiura, M, Ibaraki,T. and Glover, F. A path relinking approach for the generalized assignment problem. Proc. of the Int. Symp. on Scheduling, Japan, June 4-6, (2002) 105-108.
15. Yagiura, M, Ibaraki,T. and Glover, F. An ejection chain approach for the generalized assignment problem. INFORMS J Computing 16 (2004a) 133-151.
16. Yagiura, M, Iwasaki, S, Ibaraki, T, & Glover, F. A very large-scale neighbourhood search algorithm for the multi-resource generalised assignment problem. Discrete Optimization 1 (2004b) 87-98.

Parametrized Greedy Heuristics in Theory and Practice

Armin Fügenschuh

Darmstadt University of Technology,
Schlossgartenstr. 7, 64289 Darmstadt, Germany
`fuegenschuh@mathematik.tu-darmstadt.de`

Abstract. A parametrized greedy heuristic (pgreedy) is a hybridization of a classical greedy construction heuristic with a parametrized scoring function and improving hit-and-run, a concept from the field of Global Optimization for automated parameter selection. In this article we introduce this new metaheuristic concept and apply it to a real-world planning problem: the integrated coordination of school starting times and public transport in rural areas.

1 Parametrized Greedy Heuristics

Greedy-type heuristics are used in many special-purpose optimization software packages, where a good feasible solution to a given instance of some (combinatorial) problem is required after a very short amount of time (typically, a few seconds). They construct feasible solutions from scratch by step-by-step inserting always the best immediate, or local, solution while finding an answer to the given problem instance. To obtain in fact good solutions, the crucial point within every greedy algorithm is having a proper criterion that selects these local solutions and thus is responsible for the search direction. For some optimization problems greedy algorithms are able to find the globally optimal solution. For example, Prim's or Kruskal's algorithms actually are greedy algorithms which find a minimum spanning tree in a given graph. On the other hand, there is no known greedy algorithm that finds a minimum Hamiltonian path, i.e., a solution to the traveling salesman problem (TSP).

Consider an optimization problem in the general form

$$z^{\text{opt}} = \min\{c(x) : x \in S\}, \tag{1}$$

where $S \subset \mathbb{Z}^n$ is a non-empty, bounded (hence finite) set of feasible solutions and $c : S \to \mathbb{Q}$ is an objective function. Let $N := N_0 := \{1, \dots, n\}$. A *scoring function* is a mapping $s : N \times \mathbb{Z} \to \mathbb{Z} \cup \{\infty\}$ which yields a score $s(i, v)$ for the assignment of value $v \in \mathbb{Z}$ to variable x_i with index $i \in N$. Assume that it is computationally easy to identify some $(i, v) \in N \times \mathbb{Z}$ that minimizes the scoring function. A *greedy heuristic* is defined as a procedure that selects in the k-th step an index $i \in N_k$

M.J. Blesa et al. (Eds.): HM 2005, LNCS 3636, pp. 21–31, 2005.

and a value $v \in \mathbb{Z}$ with $(i, v) = \text{argmin}\{s(j, w) : j \in N_k, w \in \mathbb{Z}\}$ and sets $x_i := v$. Then k is increased by one and $N_k := N_{k-1}\backslash\{i\}$. The greedy algorithm terminates after at most n steps when there is no free (unassigned) variable left, i.e., $N_k = \emptyset$. As one can imagine the actual selection of a proper scoring function s is essential for constructing a good greedy heuristic. The algorithm's output is a vector x called *solution*. A solution x is either *feasible* if $x \in S$, or infeasible otherwise. A feasible solution is called *optimal* if for every other feasible solution y we have $c(x) \leq c(y)$. If a greedy heuristic always terminates with a feasible solution it is called *reliable*. If a reliable greedy heuristic always produces optimal solutions it is called *exact*.

In case of the TSP a simple greedy heuristic is the *nearest-neighbor heuristic* (NN) that works as follows. Given is a complete graph $G = (V, A)$ with non-negative arc weights $c_{vw} \in \mathbb{Q}_+$. The salesman starts at an arbitrary node v_0 and then visits the node nearest to the starting node. That is, a node v_1 with $v_1 = \text{argmin}\{s_{v_0 w} : (v_0, w) \in A\}$ is selected, where the scoring function s is defined as

$$s_{vw} := c_{vw}, \quad \forall\, (v, w) \in A. \tag{2}$$

From there the salesman visits the nearest node v_2 that was not visited so far. In general, in the k-th step of the heuristic we seek

$$v_k = \text{argmin}\{s_{v_{k-1}w} : w \in V, (v_{k-1}, w) \in A_k\}, \tag{3}$$

with $A_k := A_{k-1}\backslash\{(v_{k-1}, v_0), \ldots, (v_{k-1}, v_{k-2})\}$ for all $k \geq 2$ and $A_1 := A$, and insert it into the tour. These steps are iteratively repeated until all nodes are visited and the salesman returns to the start node v_0. This maybe is the first heuristic that almost everyone comes up with. It is probably close to a practitioner's approach. However, the solutions found by this heuristic sometimes are of poor quality. The reason is, as one can see in Figure 1 (left; tour length 677 units) below, that some nodes are "forgotten" during the course of the algorithm. They have to be inserted in later steps towards the end at relatively high costs.

A *parametrized scoring function* with p parameters $\lambda \in \mathbb{Q}^p$ is a mapping $s : N \times \mathbb{Z} \times \mathbb{Q}^p \to \mathbb{Q}$ with argument (i, v, λ). In the sequel we restrict our discussion to the case of scoring functions that are linear in the parameters, i.e., for all $i \in N, v \in \mathbb{Z}, \lambda, \mu \in \mathbb{Q}^p$, and $t, u \in \mathbb{Q}$ we have $s(i, v, t \cdot \lambda + u \cdot \mu) = t \cdot s(i, v, \lambda) + u \cdot s(i, v, \mu)$. A greedy heuristic that makes use of such a scoring function is hence called a *parametrized greedy heuristic* or *pgreedy*, for short. The local selection of variables and values and hence the entire solution found by the pgreedy heuristic depends on the actual choice of $\lambda \in \mathbb{Q}^p$. We write $x(\lambda)$ for the solution found when the pgreedy algorithm is called with parameter λ, and $z(\lambda)$ for the corresponding objective function value, i.e., $z(\lambda) = c(x(\lambda))$. We are now faced with the problem to find a vector λ with $z(\lambda) \leq z(\mu)$ for all $\mu \in \mathbb{Q}^p$ and hence to search for

$$z^{\text{pgreedy}} = \inf\{z(\lambda) : \lambda \in \mathbb{Q}^p\}. \tag{4}$$

If the domain for the parameters is bounded – and we show in Corollary 1 below that this can be assumed without loss of generality – then in principle the optimal parameter vector λ can be found by sampling over a regular, sufficiently dense grid. For each λ the pgreedy heuristic has to be called and the best λ (i.e., the λ with the lowest objective function value $z(\lambda)$) is kept. However, in practice this approach turns out to be inefficient, even for a relative few number of parameters. Another idea is to select candidate λ parameters randomly, which also leads to a high number of runs.

Instead of those we use *improving hit-and-run* (*IHR* for short), an algorithm introduced by Zabinsky et al. [7] (see also [6]) to solve general global optimization problem. IHR is a randomized (Monte-Carlo) algorithm that automatically selects parameters which lead to good, possibly optimal solutions when used in a pgreedy heuristic. In a hybrid algorithm of pgreedy and IHR, a combination of a parametrized greedy algorithm and improving hit-and-run, IHR is used to compute the weights λ that take control of the parametrized scoring function and calls the pgreedy algorithm as a black-box to obtain a new objective function value. The basic idea behind improving hit-and-run is to use hit-and-run to generate candidate points randomly and accept only those that are an improvement with respect to the objective function. For a given objective function $z : B \to \mathbb{Q}$, defined on a bounded subset $B \subset \mathbb{Q}^p$ the IHR algorithm works as follows: We start with an initial $\lambda^0 \in B$, and set $k := 0$. The following steps are now repeated until a stopping criterion is met, for example, if the number of iterations k reaches a certain limit. Generate a random direction vector d_k uniformly distributed on the boundary of the unit hypersphere $\mathbb{S}^p := \{\lambda \in \mathbb{Q}^p : \|\lambda\|_2 = 1\}$. Generate a candidate point $w^{k+1} := \lambda^k + t \cdot d^k$, where t is generated by sampling uniformly over the line set $L_k := \{\lambda \in B : \lambda = \lambda^k + t \cdot d^k, t \in \mathbb{Q}_+\}$. If the candidate point is improving, i.e., $z(w^{k+1}) < z(\lambda^k)$, we set $\lambda^{k+1} := w^k$, otherwise $\lambda^{k+1} := \lambda^k$. Finally, increase k by 1.

However, the domain of parameters for the objective function has to be bounded to apply IHR. The subsequent Theorem 1 and the derived Corollary 1 are crucial to restrict the unbounded domain for the λ parameters to some bounded subset $B \subset \mathbb{Q}^p$. Without the assertion of the Corollary, IHR would not be applicable.

Theorem 1. *Let* $\lambda, \lambda' \in \mathbb{Q}^p$. *If there is a positive scalar* $t \in \mathbb{Q}_+$ *such that* $\lambda' = t \cdot \lambda$ *then* $x(\lambda) = x(\lambda')$ *and hence* $z(\lambda) = z(\lambda')$.

Proof. Consider an arbitrary step k of the pgreedy heuristic with parameter λ. Let $(i, v) := \operatorname{argmin}\{s(j, w, \lambda) : j \in N, w \in \mathbb{Q}\}$ be the selected index and value in this step. That means, $s(i, v, \lambda) \leq s(j, w, \lambda)$ for all $j \in N, w \in \mathbb{Q}$. It follows that $t \cdot s(i, v, \lambda) \leq t \cdot s(j, w, \lambda)$. Since $s(i, v, \cdot)$ is linear, we get $s(i, v, t \cdot \lambda) \leq s(j, w, t \cdot \lambda)$ and hence $s(i, j, \lambda') \leq s(j, w, \lambda')$ for all $j \in N, w \in \mathbb{Q}$. Therefore $(i, v) = \operatorname{argmin}\{s(j, w, \lambda') : j \in N, w \in \mathbb{Q}\}$. Since k was arbitrarily chosen both heuristics select the same local best node in every step. Thus the constructed solutions and their objective function values are the same.

Corollary 1. *Let* $\| \cdot \|$ *be an arbitrary norm. For every solution* $x(\lambda)$ *with* $\lambda \in \mathbb{Q}^p \backslash \{0\}$ *there is a* $\lambda' \in \mathbb{Q}^p$ *with* $\|\lambda'\| = 1$ *such that* $x(\lambda) = x(\lambda')$.

Proof. Let $\lambda \in \mathbb{Q}^p, \lambda \neq 0$. If already $\|\lambda\| = 1$ then we have nothing to do anymore. Otherwise set $t := 1/\|\lambda\|$ and $\lambda' := t \cdot \lambda$. Then $\|\lambda'\| = 1$ and $x(\lambda) = x(\lambda')$ by Theorem 1.

In particular, we have shown that

$$z^{\text{pgreedy}} = \inf\{z(\lambda) : \lambda \in \mathbb{Q}^p\} = \min\{z(\lambda) : \|\lambda\| = 1\}. \tag{5}$$

In particular this means that the search for parameters λ can be restricted to the unit hypersphere (instead the whole $\mathbb{Q}^p$), which considerably speeds up computations.

We now continue the TSP example and extend the nearest neighborhood heuristic to a parametrized greedy heuristic. The crucial point for pgreedy is that more than a single criterion is needed. Finding those additional criteria is in general more of an art than science. For example, in the k-th step of the nearest-neighbor heuristic (with $1 \leq k \leq |V|$) one can consider the "way back to the start node", i.e., $c_{v_k v_0}$ as an additional term in the scoring function. Hence we obtain

$$s_{vw}(\lambda_1, \lambda_2) := \lambda_1 \cdot c_{vw} + \lambda_2 \cdot c_{wv_0}, \quad \forall\, (v, w) \in A, \tag{6}$$

as parametrized scoring function, where λ_1 and λ_2 are some scalar parameters. After a few iterations of the IHR algorithm a better solution for this TSP instance is obtained with parameters $\lambda_1 := 0.6$ and $\lambda_2 := -0.4$ in the scoring function (6). This solution is shown in Figure 1 (middle; tour length 592 units).

However, the generated solution still can be improved. This is done by a final local search step. A local search strategy for TSP is the so-called k-opt. A k-opt is an exchange of $k \in \mathbb{Z}, k \geq 2$ arcs of the solution. A recombination of this kind is accepted if the overall tour length is reduced, and rejected otherwise. k-opt steps are sequentially applied until no improvement is found anymore. Figure 1 (right; tour length 525 units) shows the solution that is found for $k = 2$.

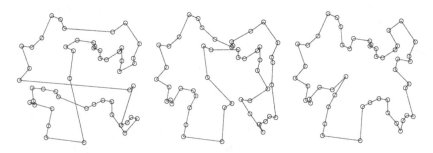

Fig. 1. Greedy (677), pgreedy (592), and pgreedy with local search (525)

These three components, a greedy heuristic using a parametrized scoring function, a meta-algorithm in which this greedy heuristic is embedded to find the "right" parameters, and a final local search procedure, are the main ingredients for a successful pgreedy heuristic.

2   A Real-World Planning Problem

Originally pgreedy was developed to aid human planners solving the integrated planning problem of coordinating bus and school starting times and bus schedules (the project acronym is IKOSANA, an abbreviation of the German "<u>I</u>ntegrierte <u>Ko</u>ordinierung von <u>S</u>chul<u>a</u>nfangszeiten und des <u>Na</u>hverkehrs-<u>A</u>ngebots"). This problem occurs frequently in every county in Germany, and with its solution large amounts of money can be saved.

Let $\mathcal{V}$ be the set of all passenger trips in the given county. A *trip* $t \in \mathcal{V}$ is a sequence of bus stops, each having an arrival and a departure time assigned to. The time difference δ_t^{trip} between the departure at the first and the arrival at the last bus stop is called the *service duration*. (All time-related parameters and variables in the model are integral with the unit "minute".) For every trip $t \in \mathcal{V}$ we introduce an integer variable $\alpha_t \in \mathbb{Z}_+$ representing its planned starting time, i.e., the departure of a bus at the first bus stop. A time window $\underline{\alpha}_t, \overline{\alpha}_t$ is given, in which the planned trip starting time must be:

$$\underline{\alpha}_t \leq \alpha_t \leq \overline{\alpha}_t. \tag{7}$$

The trips are served by buses which start and end their services at a *depot*. In principle, every trip may be served by a new bus from the depot. The trip without passengers from the depot to the first bus stop of trip t is called *pull-out trip*. When the bus arrives at the last bus stop of passenger trip t, it is either sent on the *pull-in trip*, i.e., back to the depot, or it is re-used to serve another passenger trip. The duration of the pull-out and pull-in trips is denoted by $\delta_t^{\text{out}}, \delta_t^{\text{in}}$, respectively.

Instead of sending the bus back to the depot after having served a trip it is of course more sensible to re-use the bus to serve other trips, as long as this is possible. Thus we seek such a connection of trips. Let the set $\mathcal{A} \subset \mathcal{V} \times \mathcal{V}$ contain all pairs of trips (t_1, t_2) that can in principle be connected. The intermediate trip from the last bus stop of trip t_1 to the first bus stop of trip t_2, where no passengers are transported, is called a *shift* or a *deadhead trip*, its duration is $\delta_{t_1 t_2}^{\text{shift}}$. The connection of a pull-out trip, passenger and deadhead trips and a final pull-in trip which are to be served by one and the same bus is called a *block* or *schedule*. For every trip $t \in \mathcal{V}$ the decision variables $v_t, w_t \in \{0, 1\}$ indicate if trip t is the first or the last trip in some block, respectively. For every pair of trips $(t_1, t_2) \in \mathcal{A}$ the variable $x_{t_1 t_2} \in \{0, 1\}$ indicates if t_1 and t_2 are in sequence in some block, that is, the same bus serves trip t_2 directly after trip t_1.

Each trip is served by exactly one bus. That means, trip $t_2 \in \mathcal{V}$ either a unique predecessor or it is the first one in some block:

$$\sum_{t_1 : (t_1, t_2) \in \mathcal{A}} x_{t_1 t_2} + v_{t_2} = 1. \tag{8}$$

Moreover, every trip $t_1 \in \mathcal{V}$ either has a unique successor or it is the last one in some block:

$$\sum_{t_2 : (t_1, t_2) \in \mathcal{A}} x_{t_1 t_2} + w_{t_1} = 1. \tag{9}$$

If trips $(t_1, t_2) \in \mathcal{A}$ are connected, then trip t_2 can only start after the bus has finished trip t_1, shifted from the end of t_1 to the start of t_2, and has waited a specified time to absorb possible delays. Additional waiting is permitted within certain limits if the bus arrives before the start of t_2. Using a sufficiently big value for M, these constraints can be formulated as linear inequalities:

$$\alpha_{t_1} + \delta_{t_1}^{\text{trip}} + \delta_{t_1 t_2}^{\text{shift}} - M \cdot (1 - x_{t_1 t_2}) \leq \alpha_{t_2}, \tag{10}$$

where M is a large but finite number.

Let $\mathcal{S}$ be the set of all schools in the county under consideration. It is allowed to change this starting time within some time window (by law it is allowed that schools start between 7:30 and 8:30 a.m.). Moreover the school starting time is required to be in discrete time slots of 5 minutes (7:30, 7:35, 7:40, etc.). For every school $s \in \mathcal{S}$ a time window $\underline{\tau}_s, \overline{\tau}_s$ is given and we introduce an integer variable $\tau_s \in \mathbb{Z}_+$ with

$$\underline{\tau}_s \leq 5 \cdot \tau_s \leq \overline{\tau}_s. \tag{11}$$

Thus the planned school starting time of s is given by $5 \cdot \tau_s$.

The set $\mathcal{P} \subset \mathcal{S} \times \mathcal{V}$ consists of pairs (s, t), where trip t transports pupils to a bus stop of school s. In this case we say, t is a *school trip* for s. The time difference between the departure at the first bus stop of t and the arrival at the bus stop of s is $\delta_{st}^{\text{school}}$. There is another time window for the pupils $\underline{\omega}_{st}^{\text{school}}, \overline{\omega}_{st}^{\text{school}}$, specifying the minimal and maximal waiting time relative to the school starting time. The lower bound $\underline{\omega}_{st}^{\text{school}}$ is chosen according to the walking time from the bus stop where the pupils are dropped off, whereas the upper bound $\overline{\omega}_{st}^{\text{school}}$ is due to law restrictions. A typical time window is $5 - 45$ minutes. For every $(s, t) \in \mathcal{P}$ the starting times of trip t and school s have to be chosen such that the waiting time restrictions for the pupils at school s are met. Thus, we add the following inequalities to the model in order to synchronize the start of bus trips and schools:

$$\begin{aligned} \alpha_t + \delta_{st}^{\text{school}} + \underline{\omega}_{st}^{\text{school}} &\leq 5 \cdot \tau_s, \\ \alpha_t + \delta_{st}^{\text{school}} + \overline{\omega}_{st}^{\text{school}} &\geq 5 \cdot \tau_s. \end{aligned} \tag{12}$$

The optimization basically has two goals: Finding a minimum number of buses to serve all trips, and deploying these buses in the most efficient way, that is, minimize the sum of all deadhead trips (including pull-out, pull-in trips). As objective function we thus have

$$M_1 \cdot \sum_{t \in \mathcal{V}} v_t + M_2 \cdot \left(\sum_{t \in \mathcal{V}} \delta_t^{\text{out}} \cdot v_t + \sum_{(t_1, t_2) \in \mathcal{A}} \delta_{t_1 t_2}^{\text{shift}} \cdot x_{t_1 t_2} + \sum_{t \in \mathcal{V}} \delta_t^{\text{in}} \cdot w_t \right), \tag{13}$$

where again M is a sufficiently big value such that the reduction of the fleet size dominates over the length of the deadhead trips.

Summing up we want to solve the following integer optimization problem (also called IKOSANA in the sequel):

$$\begin{aligned} &\min (13) \\ &\text{subject to } (7), \ldots, (12) \\ &v, w \in \{0, 1\}^{|\mathcal{V}|}, x \in \{0, 1\}^{|\mathcal{A}|}, \tau \in \mathbb{Z}^{|\mathcal{S}|}, \alpha \in \mathbb{Z}^{|\mathcal{V}|}. \end{aligned} \tag{14}$$

3   Solving IKOSANA with PGreedy

From a theoretical point of view, we remark that IKOSANA is a generalization of the vehicle routing problem with time windows (VRPTW), a problem known to be theoretically and practically difficult (for a proof of the NP-completeness see Savelsbergh [4], for a general introduction to vehicle routing see Toth and Vigo [5]). The new aspect in addition to VRPTW are the coupling conditions on the starting time windows given by inequalities (12). (We suggest the term *vehicle routing problem with coupled time windows* (VRPCTW) for this new problem class.) Hence from a theoretical point of view, solving IKOSANA to optimality is NP-hard, since checking feasibility with a fixed number of vehicles is NP-complete. Thus we cannot expect a polynomial algorithm for its solution unless $P = NP$. In order to obtain good feasible solutions we now describe an application of the parametrized greedy heuristic to IKOSANA. The entire solution process is split into two stages. In stage one schedules of the buses are computed such that preferably few buses are in use and the total length of all deadhead-trips is minimal. At this stage, it is not necessary to compute starting times for the schools and the trips. It only has to be assured that the schedules of the buses are feasible in such way that feasible school starting times can be computed afterwards. Their actual computation is then done in stage two where the schedules from stage one are taken as fixed input values.

3.1   Starting Time Propagation

The most important subroutine within the heuristic is a constraint propagation algorithm for bounds strengthening. Note that the subsystem of the IKOSANA model (14) consisting of inequalities (7), (10) for those inequalities with $x_{t_1 t_2} = 1$, (11), and (12) is a *monotone IP2*, i.e., an integer program where each inequalities has at most two non-zero coefficients (IP2) and, if there are exactly two non-zeros, one is positive and the other is negative (monotone). Finding feasible solutions of a monotone IP2 is also an NP-hard problem, since it contains simultaneous Diophantine approximation as a special case (for the details we refer to Lagarias [2]). For the construction of feasible solutions of these systems, or to show that no such solution exists, Bar-Yehuda and Rawitz [1] proposed the following pseudo-polynomial algorithm.

Given are the integer variables $y_i \in \mathbb{Z}$ for $i = 1, \ldots, n$ (which now represent the α and τ variables in the IKOSANA model) with bounds $l_i \leq y_i \leq u_i$ and the monotone inequality system $a_k y_{i_k} + b_k y_{j_k} \leq c_k$ for $k = 1, \ldots, m$, with $a_k \cdot b_k < 0$. The main idea of the algorithm is to use constraint propagation to find the impact of constraint k and the bounds l_{i_k}, u_{i_k} on the bounds l_{j_k}, u_{j_k}. If $a_k > 0$ and $b_k < 0$ we obtain a new lower bound on y_{j_k} by setting

$$l_{j_k} := \max \left\{ l_{j_k}, \left\lceil \frac{c_k - a_k l_{i_k}}{b_k} \right\rceil \right\}. \tag{15}$$

Otherwise, if $a_k < 0$ and $b_k > 0$, a new upper bound on y_{j_k} is given by:

$$u_{j_k} := \min \left\{ u_{j_k}, \left\lfloor \frac{c_k - a_k u_{i_k}}{b_k} \right\rfloor \right\}. \tag{16}$$

(For the impact of the bounds l_{j_k}, u_{j_k} on the bounds l_{i_k}, u_{i_k} similar formulas can be derived.) These two propagation steps (15) and (16) are now iteratively repeated for all variables, until no bound is improved any more. If $u_{j_k} < l_{j_k}$ occurs after the update then the subsystem is infeasible. Bar-Yehuda and Rawitz showed that the entire propagation can be implemented such that its time complexity is $O(mU)$, with $U := \max\{u_i - l_i : i = 1, \ldots, n\}$.

This bounds strengthening algorithm is used to evaluate the influence of one starting time window (of either a trip or a school) on all other time windows. Thus we call it *starting time propagation* in the sequel. Starting time propagation is used in the beginning to check whether the given instance is feasible at all. It is then used after every construction step in the pgreedy heuristic to check whether a connection of trips is feasible and to assure that starting times of schools and trips can be computed in stage two of the heuristic.

3.2 Constructing the Schedules

The construction of feasible schedules for all vehicles is the first stage of our heuristic, after the instance is checked to be feasible. Consider the graph $(\mathcal{V}, \mathcal{A})$. In the k-th step of the heuristic a local-best deadhead-trip from $\mathcal{A}$ is selected. Before we remove those arcs from $\mathcal{A}$ that would lead to infeasibilities when being chosen. An arc $(t_1, t_2) \in \mathcal{A}$ is infeasible in view of inequalities (10) if the vehicle arrives at the first bus stop of t_2 after the latest possible start of t_2. Another kind of infeasibility occurs in view of inequalities (8) and (9). If some $x_{t_1 t_2}$ is already fixed to its upper bound 1 then all arcs $(t, t_2), (t_1, t) \in \mathcal{A}$ can also be removed from $\mathcal{A}$ since every tour must have a unique predecessor and successor, respectively. The same applies for trips that are already selected as first or last trip in some schedule. In order to remove all those arcs that would make the solution infeasible when being selected we define $\mathcal{A}_k \subseteq \mathcal{A}$ as

$$
\begin{aligned}
\mathcal{A}_k := \mathcal{A}_{k-1} \backslash \{ \ & \{(t_1, t_2) \in \mathcal{A} : \underline{\alpha}_{t_1} + \delta_{t_1}^{\text{trip}} + \delta_{t_1 t_2}^{\text{shift}} > \overline{\alpha}_{t_2}\} \\
\cup \ & \{(t_1, t_2) \in \mathcal{A} : \exists t \in \mathcal{V}, x_{t_1 t} = 1\} \\
\cup \ & \{(t_1, t_2) \in \mathcal{A} : \exists t \in \mathcal{V}, x_{t t_2} = 1\} \\
\cup \ & \{(t_1, t_2) \in \mathcal{A} : w_{t_1} = 1\} \\
\cup \ & \{(t_1, t_2) \in \mathcal{A} : v_{t_2} = 1\} \ \}
\end{aligned}
\tag{17}
$$

for all $k \geq 1$ and $\mathcal{A}_0 := \mathcal{A}$. We set $x_{t_1 t_2} := 0$ for all removed arcs $(t_1, t_2) \in \mathcal{A}_{k-1} \backslash \mathcal{A}_k$.

For the scoring function we discuss two alternatives. Similar to the introductory nearest neighbor TSP example we take

$$
s_{t_1 t_2} := \delta_{t_1 t_2}^{\text{shift}},
\tag{18}
$$

as scoring function. Hence an arc $(t_1, t_2) \in \mathcal{A}_k$ with minimal distance $\delta_{t_1 t_2}^{\text{shift}}$ is selected, i.e.,

$$
(t_1, t_2) = \text{argmin}\{s_{ij} : (i, j) \in \mathcal{A}_k\}.
\tag{19}
$$

However, this scoring function (18) yields very poor solutions with a high number of vehicles, because it is by definition "blind" for everything that has to do with time windows. In IKOSANA the connection of two trips t_1 and t_2 does affect some or all of the time windows. Vice versa, the time windows affect the set of deadhead trips that remain to be selected in the next round of the heuristic. Thus, we seek a scoring function that does not only take into account the time for the deadhead trip, but also takes care of the necessary changes in the corresponding time windows. If the time windows are narrowed too early in the course of the algorithms, the number of deployed buses quickly increases, because no "flexibility" remains. Thus, we introduce a scoring function that prefers those connections that do not (or at least not too much) change time windows of other trips or schools. For this, we define

$$
\begin{aligned}
s_{t_1 t_2}(\lambda) := \quad & \lambda_1 \cdot \delta_{t_1 t_2}^{\text{shift}} \\
+ \ & \lambda_2 \cdot |\underline{\alpha}_{t_1} + \delta_{t_1}^{\text{trip}} + \delta_{t_1 t_2}^{\text{shift}} - \underline{\alpha}_{t_2}| \\
+ \ & \lambda_3 \cdot |\overline{\alpha}_{t_1} + \delta_{t_1}^{\text{trip}} + \delta_{t_1 t_2}^{\text{shift}} - \overline{\alpha}_{t_2}| \\
+ \ & \lambda_4 \cdot |\underline{\alpha}_{t_1} + \delta_{t_1}^{\text{trip}} + \delta_{t_1 t_2}^{\text{shift}} - \overline{\alpha}_{t_2}| \\
+ \ & \lambda_5 \cdot |\overline{\alpha}_{t_1} + \delta_{t_1}^{\text{trip}} + \delta_{t_1 t_2}^{\text{shift}} - \underline{\alpha}_{t_2}|
\end{aligned}
\tag{20}
$$

with $\lambda = (\lambda_1, \ldots, \lambda_5) \in \mathbb{Q}_+^5$. The search of a good set of parameters $\lambda \in \mathbb{Q}_+^5$ is handed over to improving hit-and-run. Since (20) is linear in the parameters Corollary 1 applies to our situation. Even more, in our scoring function the λ parameters are non-negative. Thus we can restrict the domain of parameters to the standard simplex $\{\lambda \in \mathbb{Q}_+^5 : \lambda_1 + \ldots + \lambda_5 = 1\}$.

For the arc (t_1, t_2) found in (19) we now set $x_{t_1 t_2} := 1$. Due to the connection of trip t_1 with t_2 the starting time windows of these two trips might change. This of course propagates onto all other starting time windows of trips and schools. To evaluate the effect of the connection we call the starting time propagation again.

The above steps are repeated iteratively until $\mathcal{A}_k = \emptyset$ for some k. This is equivalent to the fact that all binary variables x are fixed to their bounds, 0 or 1. Since v and w are slack variables their values are automatically determined by the x variables. In an intermediate step, we improve the solution by a 2-opt local search step. Thus, we end up with a schedule for all buses and strengthened bounds on the time variables, which is now input for the second stage of the optimization, the starting time computation.

3.3   Assigning the Starting Times

The goal of stage two is the computation of school and trip starting times, given the bus schedules from stage one. During the computations in stage one, the time windows have been significantly narrowed. If a time window shrinks to a single point, then the corresponding variable, α_t or τ_s, can be fixed to this value. However, for the majority of all time windows, there is still some flexibility.

We start with the schools and distinguish three cases. If the former school starting time $\hat{\tau}_s$ of school $s \in \mathcal{S}$ is below the lower bound $\underline{\tau}_s$, we fix the school starting time to the lower bound, that is, we let $\tau_s = \frac{1}{5}\underline{\tau}_s$, and the school s starts at $\underline{\tau}_s$. If the former school starting of school s is above the upper bound $\overline{\tau}_s$, we fix the school starting time to the upper bound, that is, we let $\tau_s = \frac{1}{5}\overline{\tau}_s$, and the school s will start at $\overline{\tau}_s$. Finally, if the former school starting of school s is between the upper and the lower bound, $\underline{\tau}_s \leq \hat{\tau}_s \leq \overline{\tau}_s$, then this starting time is kept, i.e., we let $\tau_s = \frac{1}{5}\hat{\tau}_s$ and school s is going to start at $\hat{\tau}_s$. The idea behind this settings is to keep the new school starting times as close as possible to the former starting times that the pupils and teachers are already custom to. After each fixing of a school starting time, another starting time propagation step has to be carried out to evaluate the impact of the respective setting for other school and trip starting times.

After having settled the school starting times, we assign starting times to the trips in the very same fashion. Thus when the heuristic terminates after the second stage a feasible solution x, v, w, τ, α to the problem is returned.

4   Input Data and Computational Results

The IKOSANA pgreedy heuristic was implemented in C++ on a 2.6GHz Intel-Pentium IV computer running debian Linux and tested using five real-world data sets. County ct_1 is located in the state Mecklenburg-Western Pomerania, counties ct_2, ct_3, and ct_5 are all in North Rhine-Westphalia, and county ct_4 is in Saxony-Anhalt. The sizes of sets in instances ct_1 to ct_5 are shown in Table 1.

Table 1. Size of the input data sets

instance	$|\mathcal{V}|$	$|\mathcal{A}|$	$|\mathcal{S}|$	$|\mathcal{P}|$
ct_1	247	60,762	43	195
ct_2	490	239,610	102	574
ct_3	191	36,290	82	294
ct_4	134	17,822	37	201
ct_5	404	162,812	84	579

Table 2. Computational results

instance	current		g	time	pg-rand	pg-ihr	iter.	CPlex
ct_1	82.0423	81.0358	6	65.0953	65.0882	2,300	66.0553
ct_2	226.0711	227.0556	327	173.2122	175.1595	120	187.1361
ct_3	90.0488	83.0410	11	66.1225	66.1106	2,500	70.0864
ct_4	43.0093	49.0138	1	38.0407	38.0268	7,500	38.0198
ct_5	176.0725	180.0633	37	135.1790	134.1704	900	140.1084

In Table 2 we summarize our computational results. In the first column is the number of currently used buses and the length of all deadhead, pull-in and pull-out trips. (Since buses are more expensive than deadhead trips we use $M_1 := 1$ and $M_2 := \frac{1}{10000}$. For example, in county ct_1 there are currently 82 buses that need 423 minutes for all kind of deadhead trips.) The solutions found when using the greedy scoring function (18) are shown in column "g". It turns out that the nearest neighbor insertion is a bad strategy: In all but one case the number of buses is even higher after the optimization than before. The time for a single call of the greedy heuristic is given in the left column "time". The computation time mainly depends on the number of deadhead trips within the respective instance, and ranges from less than a second to a few minutes. In column "pg-rand" we show the results when sampling parameters randomly for the parametrized scoring function (20). In column "pg-ihr" improving hit-and-run was applied instead of pure random sampling. In all runs we limited the computation time to 3,600 seconds, which gives a different number of iterations (calls of the greedy heuristic) for each instance (see column "iter"). Compared to random parameter selection it turns out that in all but one case IHR leads to better results after the same number of iterations. We also solved (14) using a commercial state-of-the-art MIP solver [3], see column "CPlex". It turns out that in all but one case both pgreedy heuristics (with random and IHR parameter selection strategy) found better feasible solutions within the given time limit (3,600 seconds).

Compared to the currently used number of buses we are able to detect savings of 15 − 30%. The reduction of a single bus yields savings of 35,000 Euro per year for the county's tax payers. In practice the integrated coordination of school starting times is just at the very beginning. One can safely assume that in most counties in Germany the school starting times are not (or at least not sufficiently) coordinated with public transport. Extrapolating this savings to all 323 counties in Germany, we come up with savings of 200 − 300 Mio. Euro every year. So, what are we waiting for?

References

1. Bar-Yehuda R., Rawitz D. (2001), Efficient algorithms for integer programs with two variables per constraint. *Algorithmica* **29** (4), 595 − 609.
2. Lagarias J.C. (1985), The computational complexity of simultaneous diophantine approximation problems. *SIAM Journal on Computing* **14**, 196 − 209.
3. ILOG CPLEX Division, 889 Alder Avenue, Suite 200, Incline Village, NV 89451, USA. Information available at URL http://www.cplex.com.
4. Savelsbergh M. (1986), Local Search for Routing Problems with Time Windows. *Annals of Operations Research* **4**, 285 − 305.
5. Toth P., Vigo D. (2002), The Vehicle Routing Problem. SIAM Monographs on Discrete Mathematics and Applications. SIAM, Philadelphia.
6. Zabinsky Z.B. (2003), *Stochastic Adaptive Search for Global Optimization. Nonconvex Optimization and its Applications.* Kluwer Academic Publishers, Boston.
7. Zabinsky Z.B., Smith R.L., McDonald J.F., Romeijn H.E., Kaufman D.E. (1993), Improving Hit-and-Run for Global Optimization. *Journal of Global Optimization* **3**, 171 − 192.

A Taxonomy of Cooperative Search Algorithms

Mohammed El-Abd and Mohamed Kamel

Dept. of Electrical and Computer Engineering, University of Waterloo,
Waterloo, N2L 3G1, Canada
{mhelabd, mkamel}@pami.uwaterloo.ca

Abstract. A lot of heuristic approaches have been explored in the last
two decades in order to tackle large size optimization problems. These
areas include parallel meta-heuristics, hybrid meta-heuristic, and cooperative search algorithms. Different taxonomies have been proposed in the
literature for parallel and hybrid meta-heuristics. In these taxonomies,
one can realize that cooperative search algorithms lie somewhere in between. This paper looks at cooperative search algorithms as a stand alone
area. Two different taxonomies of cooperative search algorithm are proposed based on two different criteria. Different implementations in this
area are reported and classified using these taxonomies.

1 Introduction

Search techniques have been widely used to solve many optimization problems
of both theoretical and practical importance. They can be divided into complete (exact) and approximate (heuristic) algorithms. Complete search algorithms such as branch and bound or dynamic programming guarantee to find
an optimal solution for a finite sized problem in bounded time. However, as the
size of the problem gets larger, the time needed by the complete algorithms may
increase exponentially. On the other hand, approximate search algorithms find
a good (non-optimal) solution in less amount of time.

Another approach that was taken to solve optimization problems was the
introduction of hybrid algorithms. A hybrid algorithm is a combination of complete or approximate algorithms (or both) used to solve the problem in hand.
The interest among researchers in this field has risen in the past years since
many of the best results obtained for many optimization problems where found
by hybrid algorithms.

As the size the optimization problems that need to be solved gets larger,
researchers have tried to find a way to address these problems efficiently. The
solution to this problem was the application of parallel computation. Many parallel implementations of different search algorithms have been reported in the
literature in order to achieve a substantial decrease in the amount of time needed
to solve the problem.

Cooperative search is a category of parallel algorithms, in which several
search algorithms are run in parallel in order to solve the optimization problem in hand. The search algorithms (run in parallel) may be different, that is
why a cooperative search technique may be also viewed as a hybrid algorithm.

M.J. Blesa et al. (Eds.): HM 2005, LNCS 3636, pp. 32–41, 2005.

Sections 2 and 3 give a brief background about different taxonomies that were proposed for both hybrid and parallel algorithms respectively. Cooperative search area is identified in these taxonomies. Two different taxonomies of cooperative search algorithms are presented in section 4 along with many definitions found in the literature for these techniques. Different implementations of this technique are reported and classified in section 5. Section 6 concludes this paper.

2   Hybrid Algorithms

A taxonomy of hybrid algorithms was introduced in [20]. It was based on design and implementation issues. The former issue concerns with the architecture of the algorithms, while the later one is concerned with the hardware platform, the programming model, and the environment on which the algorithm is to run.

The taxonomy (at the design level) had both hierarchical and flat portions. Figure 1 shows the hierarchical portion with the cooperative part identified.

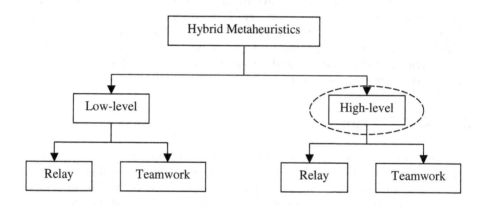

Fig. 1. Hybrid algorithms taxonomy

The cooperative part is identified in the high-level technique. In high-level relay technique, self-contained algorithms are run in a sequential manner. This technique relies on the fact that evolutionary algorithms (such as Genetic Algorithms) are good in locating promising regions in the search space, but are not suitable for fine-tuning near optimal solutions.

In this category, a local search algorithm is usually applied first to get promising initial solutions, then an evolutionary algorithm is applied to enrich these solutions, finally a local search method is applied again to fine tune these solutions. Figure 2 shows the structure of this hybrid algorithm.

In high-level teamwork technique, there are several self-contained algorithms working in parallel and cooperating to find a solution. Cooperation is evident in

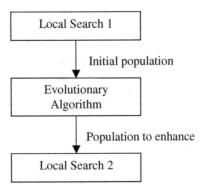

Fig. 2. High-level relay technique

this technique through the sharing of information. Note also that this technique is actually a parallel algorithm.

In [1], a different classification was given for hybrid algorithms. The authors classified the hybridization of meta-heuristics into three categories: (i) component exchange among meta-heuristics, which is similar to the low-level hybridization, (ii) cooperative search, which they relate to parallelization. So, it is similar to high-level teamwork hybridization but it is put here as a stand-alone class, (iii) integrating meta-heuristics and systematic methods. As shown, cooperative search was put in this classification as a separate class.

3 Parallel Algorithms

Great attention is paid to the area of parallel meta-heuristics in order to solve large sized optimization problems in a small amount of time. Parallel implementations of meta-heuristic algorithms give substantial decrease in time over their sequential counterparts, even getting better solutions in some situations.

The simplest classification of parallel algorithms is given in [24], parallelization of algorithms is classified into low-level and high-level parallelization.

Low-level parallelization achieves speed up by accelerating steps of the algorithm that are computationally expensive. On the other hand, high-level parallelization requires changes to be made to the original algorithm. It involves several algorithms running in parallel and cooperating by exchanging information between them. As can be seen from this classification, cooperative search algorithms are regarded as a high-level parallelization approach.

A more detailed classification [5] is illustrated in Figure 3 again with the cooperative part identified.

The three categories are: (i) Operation parallelization, which is also called low-level parallelization, (ii) Search space decomposition, in which each running algorithm searches for a sub-solution in a sub-space, these sub-solutions are used to construct a global one, which could be regarded as a cooperative scheme, and

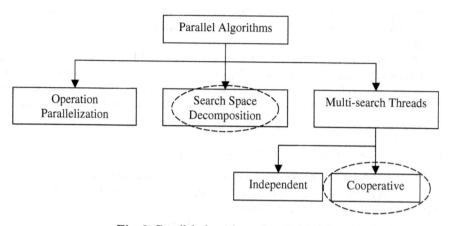

Fig. 3. Parallel algorithms classification

(iii) Multi-search threads, in which several running algorithms search for a solution in the whole space. This method could be applied with different levels of cooperation. That is why it is further divided into *independent*, no cooperation what so ever, and *cooperative*. From this classification, it can be seen that cooperative search algorithms are put as a special type of parallel algorithms.

Many different classifications, in which cooperative algorithms could be identified, have been reported as well for different meta-heuristics which include Tabu Search [6], Genetic Algorithms [4,17], and Simulated Annealing [7].

4   Cooperative Search Algorithms

4.1   Introduction

Previous discussions show that cooperative search algorithms is considered to lie somewhere between hybrid and parallel algorithms. Different classification schemes were given for both hybrid and parallel meta-heuristics. In some of them, the cooperative part is put as a stand alone class. In others, the cooperative class could be inferred.

There are many definitions in the literature for the cooperative search technique:

- "Teamwork hybridization represents cooperative optimization models, in which we have many parallel cooperating agents, each agent carries out a search in solution space" [20].
- In [21], the authors consider cooperative search as a category of parallel algorithms. "These algorithms execute in parallel several search programs on the same optimization problem instance".
- "Cooperative search algorithms are parallel search methods that combine several individual programs in a single search system" [22].

- "Cooperative search is a parallelization strategy for search algorithms where parallelism is obtained by concurrently executing several search programs" [23].
- "Cooperative search consists of a search performed by agents that exchange information about states, models, entire sub-problems, solutions or other search space characteristics" [1]. The authors refer to cooperative search as a loose form of hybridization. The authors also referred to the sequential implementation of algorithms as a loose form of cooperation.

4.2   First Taxonomy (Algorithms Based)

The first taxonomy proposed for cooperative search algorithms is based on the types of algorithms being used in the cooperative system and the implementation. Figure 4 illustrates this taxonomy.

The taxonomy shown gives rise to four different categories:

- Serial Homogenous Algorithms, this is concerned with having different instances of the same algorithms searching in different subspaces. The algorithms are run in a sequential manner, each algorithm provides a partial solution to the problem. These partial solutions are used to provide a complete solution that is evaluated and used in subsequent runs.
- Parallel Homogenous Algorithms, this category involves having different instances of the same algorithm running in parallel and searching for a solution. The information passed between these algorithms has many forms, most likely the best solution found so far. This class is identified in the hybrid taxonomy as a high-level teamwork technique.
- Serial Heterogenous Algorithms, this class involves having different algorithms running in a pipeline fashion. The output of each algorithm supplied as an input to the next algorithm. This class is identified in the hybrid taxonomy as a high-level relay technique.
- Parallel Heterogenous Algorithms, same as the second class but with different running algorithms.

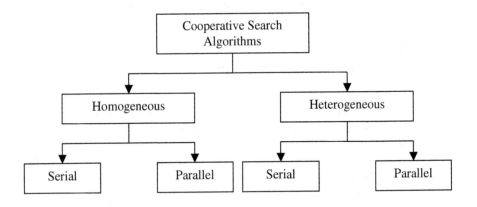

Fig. 4. Algorithms based taxonomy

4.3   Second Taxonomy (Space Decomposition Based)

Adopting a cooperative search technique always incorporates some type of space decomposition. Another taxonomy for cooperative search algorithms could be based on the type of decomposition achieved. This taxonomy is shown in figure 5.

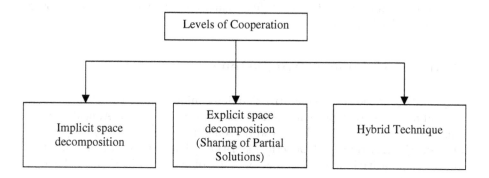

Fig. 5. Space decomposition based taxonomy

The implicit space decomposition involves the decomposition of the search space between different algorithms. This class refers to having different algorithms (or different instances of the same algorithm) looking for a solution and sharing useful information between them. There could be many choices for the information to be shared depending on the algorithms being used. The name implicit comes from the fact that the different algorithms explore different areas in the search space due to having different initial solutions, different parameter values, or both.

In the explicit space decomposition class, the search space is explicitly decomposed into sub-spaces. Each algorithm searches for a sub-solution in a different sub-space of the problem. Hence, each algorithm provides a partial solution to the problem, these partial solutions are gathered to provide a complete solution.

The hybrid approach refers to the idea of having a cooperative system that employs both methods of space decomposition.

4.4   Relating the Two Taxonomies

The two taxonomies are related to each other by the following relations:

- The implicit space decomposition class in the second taxonomy involve the implementation of parallel algorithms running concurrently and exchanging information. These algorithms could be similar or different. Hence, this class includes the parallel section of both the homogeneous and the heterogeneous classes in the first taxonomy.
- The explicit space decomposition class is similar to the homogeneous serial class in the first taxonomy. As both involve having the algorithms running in

a serial fashion providing partial solutions to the problem, these partial solutions are then gathered to generate a complete solution in order to evaluate the performance of the cooperative system.

- The hybrid class in the second taxonomy includes the same classes that are covered by the implicit space decomposition class with the difference that the algorithms now some of them explore the whole search space (providing complete solutions) and some of them search in different sub-spaces (providing partial solutions).

5 Reported Cooperative Systems

Many implementations of cooperative search techniques using Tabu Search could be found in [5], many implementation for cooperative Simulated Annealing algorithms are surveyed in [5,7]. In this work, some implementations of cooperative Tabu Search algorithms and Genetic Algorithms that were not covered are presented. Cooperative implementations for Ant Colony Optimization and Particle Swarm Optimization are also presented.

5.1 Implicit Space Decomposition Class

GAs have been widely used in cooperative environments [4]. The mostly used models are the *coarse-grained* and *fine-grained* GAs. Both techniques fall into the homogenous parallel category, the space decomposition is implicit.

In [8], a heterogenous parallel implementation which relies on implicit space decomposition was introduced. Different agents were running in parallel and exchanging information about the whole search space. Some agents used branch-and-bound and the others used GAs.

ACO [9] is an optimization technique that was inspired by the foraging behaviour of ants. Different directions have been taken to incorporate ACO in a cooperative system.

In [14] an island model which is similar to the one adopted using GAs was proposed. The cooperating colonies exchange the locally best solutions found. This implementation falls under the implicit space decomposition class.

The authors in [15,16] experimented with four different options for the information exchange between the parallel colonies in a homogenous parallel environment. These options are: (i) exchange of globally best solutions, (ii) circular exchange of locally best solutions, (iii) circular exchange of migrants, and (iv) circular exchange of locally best solutions with migrants. The authors came to the conclusion that ant algorithms with several colonies that do not exchange too much information could improve the solution quality. They also concluded that it is better to exchange the locally best solutions between neighbors in a directed ring and not too often than exchanging the local best solutions very often.

A multiple ant colony algorithm was presented in [26]. It used three ant colonies with different speed models to adaptively determine the number of clustering. Reported results on synthetic and real data showed that this approach can improve the clustering performance. This implementation is based on implicit space decomposition since the three colonies were performing the search in the whole search space. It is considered to be a homogenous parallel implementation.

The PSO [10,13] is an optimization method widely used to solve continuous nonlinear functions. It is a stochastic optimization technique that was originally developed to simulate the movement of a flock of birds or a school of fish looking for food.

A cooperative approach was introduced in [2], it was referred to as Concurrent PSO (CONPSO). This approach adopted the idea of having two swarms searching concurrently for a solution with frequent message passing of information. Hence, it relied on implicit space decomposition.

In [3], a multi-swarm technique was tested on a number of dynamic multimodal benchmark functions. A colony of multiple swarms is used. The swarms communicate with each other only when they are following attractors that are close to each other. The swarm that has a bad attractor gets all its particles positions and velocities re-initialized. The approach relied on implicit space decomposition and the swarms work in parallel. The communication is carried out synchronously after every iteration to check whether the different attractors are close to each other.

Different factors that control the behaviour of a multiple cooperating swarms approach were studied in [11,12]. These factors included: (i) the synchronization period, (ii) the communication strategy, and (iii) the number of cooperating swarms.

5.2 Explicit Space Decomposition Class

A homogenous serial implementation which relies on explicit space decomposition was reported in [19]. It was used for unconstrained non-linear function optimization. Several GAs were running in a sequential manner, each GA was concerned with optimizing a single dimension. In the end, the fitness of the complete solution is used to calculate the fitness of individuals in different populations.

A cooperative approach was introduced in [25], referred to as Cooperative PSO (CPSO). The approach relies on explicit space decomposition. It adopted the same method taken in [19] but using PSO instead of GAs. This implementation falls in the homogenous serial class.

5.3 Hybrid Class

In [23], a hierarchal cooperative model was introduced using TS. The model was applied to the graph partitioning problem. Multiple TS algorithms were working in parallel in a hierarchal fashion. Each algorithm supplies new information to the one next to it in the hierarchy. The first level in the hierarchy searches for a solution in the entire search space and it is the only level that can supply a complete solution. The other levels search for a solution in different sub-spaces. Hence, this model represents a hybrid cooperative search algorithm.

Another hybrid approach is presented in [25], referred to as hybrid CPSO (CPSO-H). The algorithm works by applying two different components in a sequential manner. One component adopts the CPSO approach, the other one is the normal PSO algorithm. Hence, one component is a cooperative system by itself adopting the explicit space decomposition approach, and the other component explores the whole search space. Each component runs only for one iteration then it passes the best solution found so far to the next component.

6   Conclusions

This paper covered many classification schemes that were proposed for hybrid and parallel meta-heuristics. The area of cooperative search algorithms are identified in the presented taxonomies. This work proposed two classification schemes for cooperative search algorithms. The classification was based on two different criteria: (i) the algorithms used, and (ii) the level of space decomposition achieved. Different implementations of cooperative search algorithms are reported and classified using these classifications.

References

1. Blum C. and Roli A. "Metaheuristics in Combinatorial Optimization: Overview and Conceptual Comparison". ACM Computing Surveys, vol. 35, no. 3, pp. 268-308, 2003.
2. Baskar S. and Suganthan P. N. "A Novel Concurrent Particle Swarm Optimization", Proceedings of the 2004 Congress on Evolutionary Computation, vol. 1, pp. 792-796, 2004.
3. Blackwell T. and Branke J., "Multi-Swarm Optimization in Dynamic Environments," in Applications of Evolutionary Computing, vol. 3005, Lecture Notes in Computer Science, G. R. Raidl, Ed.: Springer, pp. 488-599, 2004.
4. Cantu-Paz E. "A Survey of Parallel Genetic Algorithms". IllGAL Report 97003, The University of Illinois, 1997. Available on-line at: ftp://ftp-illigal.ge.uiuc.edu/pub/papers/IlliGALs/97003.ps.Z.
5. Crainic T. G. and Toulouse M. "Parallel Strategies for Metaheuristics". In F. Glover and G. Kochenberger, editors, State-of-the-Art Handbook in Metaheuristics. Kluwer Academic Publishers, 2002.
6. Crainic T. G., Toulouse M., and Gendreau M. "Towards a Taxonomy of Parallel Tabu Search Heuristics". INFORMS Journal on Computing, vol. 9, no. 1, pp. 61-72, 1997.
7. Greening D. R. "Parallel Simulated Annealing Techniques". Physica D, vol. 42, pp. 293-306, 1990.
8. Denzinger J. and Offermann T. "On Cooperation between Evolutionary Algorithms and other Search Paradigms". In Proceedings of the 1999 Congress on Evolutionary Computation, vol. 3, 1999.
9. Dorigo M., Di Caro G., and Gambardella L. M. "Ant Algorithms for Discrete Optimization". Artificial Life, vol. 5, no.2, pp. 137-172, 1999.
10. Eberhart R. C. and Kennedy J. "A New Optimizer using Particle Swarm Theory", Proceedings of the 6th International Symposium on Micro Machine and Human Science, pp. 39-43, 1995.

11. El-Abd M., Kamel M. "Multiple Cooperating Swarms for Non-Linear Function Optimization". Proceedings of the 4th IEEE International Workshop on Soft Computing for Transdiciplinary Science and Technology, 2nd International Workshop on Swarm Intelligence and Patterns, pp. 999-1008, 2005.
12. El-Abd M., Kamel M. "Factors Governing The Behaviour of Multiple Cooperating Swarms". Accepted in the Genetic and Evolutionary Computation COnference, 2005.
13. Kennedy J. and Eberhart R. C. "Particle Swarm Optimization", Proceeding of the IEEE International Conference on Neural Networks, vol. 4, pp. 1942-1948, 1995.
14. Michels R. and Middendorf M. "An Ant System for The Shortest Common Supersequence Problem". In Corne D., Dorigo M., and Glover F., editors, New Ideas in Optimization, McGrew-Hill, pp. 51-61, 1999.
15. Middendorf M., Reischle F., and Schmeck H. "Information Exchange in Multi Colony Ant Algorithms". In Rolim J., editor, Parallel and Distributed Computing, Proceedings of 15 IPDPS 2000 Workshops, 3rd Workshop on Biologically Inspired Solutions to Parallel Processing Problems, Lecture Notes in Computer Science 1800, Springer-Verlag, pp. 645-652, 2000.
16. Middendorf M., Reischle F., and Schmeck H. "Multi Colony Ant Algorithms". Journal of Heuristics, Kluwer, vol. 8, pp. 305-320, 2002.
17. Nowostawski M., and Poli R. "Parallel Genetic Algorithm Taxonomy". 3rd International Conference on Knowledge-Based Intelligent Information Engineering Systems, pp. 88-92, 1999.
18. Peram T., Veeramachaneni K., and Mohan C. K. "Distance-Fitness-Ratio Particle Swarm Optimization", Proceeding of the IEEE 2000 Swarm Intelligence Symposium, pp. 174-181, 2003.
19. Potter M .A., and de Jong K. A. "A Cooperative Coevolutinary Approach to Function Optimization", in The Third Parallel Problem Solving from Nature, Springer-Verlag, pp. 249-257, 1994.
20. Talbi E. "A Taxonomy of Hybrid Metaheuristics". Journal of Heuristics, vol. 8, no. 5, pp. 541-564, 2002.
21. Toulouse M., Crainic T. G., Sanso B., and Thulasiraman K. "Self-Organization in Cooperative Tabu Search Algorithms". IEEE International Conference on Systems, Man, and Cybernetics, vol. 3, 1998.
22. Toulouse M., Crainic T. G., and Sanso B. "An Experimental Study of The Systemic Behavior of Cooperative Search Algorithms". In Meta-Heuristics: Advances and Trends in Local Search Paradigms for Optimization. S. Voss. S. Martello, I. Osman, and C. Roucairol, editors. Kluwer Academic Publishers, pp. 373-392, 1999.
23. Toulouse M., Thulasiraman K., and Glover F. "Multi-Level Cooperative Search: A New Paradigm for Combinatorial Optimization and an Application to Graph Partitioning". In Proceeding of the 5th International Euro-Par Conference on Parallel Processing. Lecture Notes in Computer Science. Springer-Verlag, New York, pp. 533-542, 1999.
24. Trienekens H. W. J. M., and de Bruin A. "Towards a Taxonomy of Parallel Branch and Bound Algorithms". Report EUR-CS-92-01, Department of Computer Science, Erasmus University Rotterdam, 1992.
25. van den Bergh F. and Engelbrech A. P. (2004), "A Cooperative Approach to Particle Swarm Optimization", IEEE Transactions on Evolutionary Computation, vol. 8, no. 3, pp. 225-239.
26. Yang Y., and Kamel M. "Clustering Ensemble using Swarm Intelligence". In Proceedings of The 3rd Swarm Intelligence Symposium, pp. 65-71, 2003.

A Hybrid Genetic and Variable Neighborhood Descent for Probabilistic SAT Problem[*]

Zoran Ognjanović[1], Uroš Midić[1], and Nenad Mladenović[1,2]

[1] Mathematical Institute,
Kneza Mihaila 35, 11000 Belgrade, Serbia and Montenegro
{zorano, uros.m}@mi.sanu.ac.yu
[2] School of Mathematics, University of Birmingham, UK
N.Mladenovic@bham.ac.uk

Abstract. In this paper we develop a satisfiability checker for probabilistic logic. Our approach is based on a hybrid algorithm which combines genetic algorithm approach with variable neighborhood descent. Our hybrid compares favorable with previous pure genetic algorithm. Computational experiences show that problems with 200 propositional letters can be solved. They are, to the best of our knowledge, the largest PSAT-problems reported in the literature.

1 Introduction

Researchers in artificial intelligence have studied uncertain reasoning in many ways. Some of the formalisms for representing and reasoning with uncertain knowledge are based on probabilistic logic [2,10,12]. In that logic classical propositional language is expanded by expressions that speak about probability, while formulas remain true or false. It allows to make inferences in a general framework, without any special assumptions about underlying probability distributions. For example, it is possible to check consistency of rules with associated uncertainty factors and the corresponding techniques designed to handle uncertain knowledge in expert systems.

The problem of satisfiability of a probabilistic formula (PSAT, for short, also called the decision form of probabilistic satisfiability in [5]) is NP-complete [2,4]. PSAT can be reduced to the linear programming problem. However, the number of variables in the linear system corresponding to a formula is exponential in the number of propositional letters from the formula. It makes any standard linear system solving procedure (Fourier-Motzkin elimination, for example) not suitable in practice when scaling up to larger formulas. In [10] Nilsson proposed solving those large problem instances using heuristics. However, it is showed that it is still possible to use more efficient numerical methods - for example the powerful column generation procedure of linear programming [8,9].

---

[*] This research was supported by Ministarstvo nauke i zaštite životne okoline Republike Srbije, through Matematički institut, under grants 1379 and 1583.

M.J. Blesa et al. (Eds.): HM 2005, LNCS 3636, pp. 42–53, 2005.

In more recent papers [13,14], a procedure for PSAT based on the genetic algorithm (GA, for short) approach [1] was presented. Our system for PSAT was a straightforward implementation built on the top of a general GA-simulator which allowed us to concentrate on problem-specific issues, i.e., how to generate and represent problem instances, on specific GA-operators etc. Nevertheless, our satisfiability checker was able to solve, to the best of our knowledge, the largest PSAT-problems reported in the literature. We considered a set containing 27 randomly generated PSAT-instances with up to 200 propositional letters. Although the checker found solutions for some of the biggest problem instances, there were two instances which were very hard for the system which, in fact, was not able to solve them. Thus, we continued our efforts to develop an efficient and successful satisfiability checker. We built a new implementation of the checker which combines genetic algorithm approach with variable neighborhood descent (VND, for short) [6,11]. From now on, we use GA-PSAT and GA-VND-PSAT to denote our previous and the current implementation. In this paper we describe:

1. some new PSAT-specific GA-operators and the VND-scheme included in the algorithm,
2. a new, more efficient, implementation of the algorithm, and
3. some experimental analysis of its performance.

GA-VND-PSAT is a well-engineered C++ implementation of the problem-specific data structures, GA-operators, and the VND-scheme, so the qualities of the implementations of our systems are rather different. It has positive effect on the computational behavior, but new GA-operators and the VND-scheme have even more important influence on differences between GA-PSAT and GA-VND-PSAT.

In our approach, we perform the well-know translation of probabilistic formulas into linear systems. Then we use the GA-VND-approach to try to efficiently solve those systems. The price we pay for this is that we have a partial decision procedure, i.e., the algorithm cannot state that there is no solution when none is found. However, a fast algorithm may still be preferable to a slow decision procedure which guarantees to find a solution if one exists.

The rest of the paper is organized as follows. In Section 2 we give a brief description of probabilistic logic and PSAT. In Section 3 we summarize how the general GA-VND-approach is adapted to PSAT. Section 4 contains some experimental results. We give concluding remarks and directions for further investigations in Section 5.

2   Probabilistic Logic and PSAT

Let $Var = \{p, q, r, \ldots\}$ be the set of propositional letters. A weight term is an expression of the form $a_1 w(\alpha_1) + \ldots + a_n w(\alpha_n)$, where a_i's are rational numbers, and α_i's are classical propositional formulas containing propositional letters from Var. The intended meaning of $w(\alpha)$ is probability of α. A basic weight formula has the form $t \geq c$, where t is a weight term, and c is a rational number. $(t < c)$ denotes $\neg(t \geq c)$. A weight literal is an expression of the form $t \geq c$ or $t < c$. The set of all weight formulas contains all basic weight formulas, and it is closed under Boolean operations. Let α be a classical propositional formula and $\{p_1, \ldots, p_k\}$ be the set of all propositional letters that appear in α. An atom of α (also called possible world in [8,10]) is defined as a formula $at = \pm p_1 \wedge \ldots \wedge \pm p_k$ where $\pm p_i \in \{p_i, \neg p_i\}$. There are 2^k different atoms of a formula containing k primitive propositions. Let $At(\alpha)$ denote the set $\{at_1, \ldots, at_{2^k}\}$ of all atoms of α. Every classical propositional formula α is equivalent to formulas $DNF(\alpha)$ and $CDNF(\alpha) = \bigvee_{i=1}^{m} at_i$, called disjunctive normal form and complete disjunctive normal form of α, respectively. We use $at \in CDNF(\alpha)$ to denote that the atom at appears in $CDNF(\alpha)$. A formula f is in the weight conjunctive form (wfc-form) if it is a conjunction of weight literals. Every weight formula f is equivalent to a disjunctive normal form $DNF(f) = \bigvee_{i=1}^{m} \bigwedge_{j=1}^{k_i} (a_1^{i,j} w(\alpha_1^{i,j}) + \ldots + a_{n_{i,j}}^{i,j} w(\alpha_{n_{i,j}}^{i,j}) \, \rho_i \, c_{i,j})$, where disjuncts are wfc-formulas, i.e., ρ_i is either $\geq$ or $<$. Since a disjunction is satisfiable if at least one disjunct is satisfiable, we will consider formulas in the wfc-form only.

Let us suppose that we have a set T of sentences describing some events and the corresponding probabilities. Some conditions must be satisfied to guarantee consistency of T. For example, the probability of an event A must be greater than or equal to the probability of the intersection of events A and B. More formally, PSAT is the following problem: given a formula f in the wfc-form, is there any probability function defined on $At(f)$ such that f is satisfiable? Note that a wfc-formula f is satisfiable iff the following linear system is satisfiable:

$$\sum_{at \in At(f)} \mu(at) = 1$$
$$\mu(at) \geq 0, \text{ for every } at \in At(f),$$
$$(a_1 \sum_{at \in CDNF(\alpha_1)} \mu(at) + \ldots + a_n \sum_{at \in CDNF(\alpha_n)} \mu(at)) \, \rho \, c$$
$$(\text{for every weight literal } a_1 w(\alpha_1) + \ldots + a_n w(\alpha_n) \, \rho \, c \text{ in } f)$$

For example, $w(p \to q) + w(p) \geq 1.7 \wedge w(q) \geq 0.6$ is satisfiable since the same holds for the linear system

$$\mu(p \wedge q) + \mu(p \wedge \neg q) + \mu(\neg p \wedge q) + \mu(\neg p \wedge \neg q) = 1$$
$$\mu(p \wedge q) \geq 0$$
$$\mu(p \wedge \neg q) \geq 0$$
$$\mu(\neg p \wedge q) \geq 0$$
$$\mu(\neg p \wedge \neg q) \geq 0$$
$$\mu(p \wedge \neg q) + \mu(\neg p \wedge q) + \mu(\neg p \wedge \neg q) + 2\mu(p \wedge q) \geq 1.7$$
$$\mu(p \wedge q) + \mu(\neg p \wedge q) \geq 0.6.$$

NP-completeness of PSAT follows from the statement that a system of L linear (in)equalities has a nonnegative solution if it has a nonnegative solution with at most L entries positive such that the sizes of entries are bounded by a polynomial function of the size of the longest coefficient from the system [2,4].

3 GA-VND for PSAT

The input for the program is a weight formula f in the wfc-form with L weight literals. Without loss of generality, we demand that classical formulas appearing in weight terms are in disjunctive normal form. Let $Var(f) = \{p_1, \ldots, p_N\}$ denote the set of all propositional letters from f, and $|Var(f)| = N$.

An individual M consists of L pairs of the form (atom, probability) that describe a probabilistic model. The first coordinate is given as a bit string of length N, where 1 at the position i denotes $\neg p_i$, while 0 denotes p_i. Probabilities are represented by floating point numbers.

For an individual $M = ((at_1, \mu(at_1)), \ldots, (at_N, \mu(at_N)))$, the linear system is equivalent to: $\bigvee_{i=1}^{L} (\sum_{j=1}^{L} a_{ij}\mu(at_j))\, \rho_i\, c_i$. Note that it is possible that some $a_{ij} = 0$, though $[a_{ij}]$ matrix is usually not sparse.

GA-VND-PSAT has been optimized for speed (at the expense of memory space used for cached data) by caching the linear system coefficients $(a_{1j}, \ldots, a_{Lj})$ associated with each atom at_j. The coefficients are recalculated only when the atom at is changed due to mutation or crossover of the individuals.

The individuals are evaluated using function $d(M)$, which measures a degree of unsatisfiability of an individual M. Function $d(M)$ is defined as the distance between left and right hand side values of the weight literals not satisfied in the model described by M:

$$d(M) = \sqrt{\sum_{M \not\models t_i\ \rho_i\ c_i} [a_1^i \sum_{at \in \mathrm{CDNF}(\alpha_1^i)} \mu(at) + \ldots + a_{n_i}^i \sum_{at \in \mathrm{CDNF}(\alpha_{n_i}^i)} \mu(at) - c_i]^2}.$$

If $d(M) = 0$, all the inequalities in the linear system are satisfied, hence the individual M is a solution.

The following features of GA have been set for all tests. The population consists of 10 individuals; one set of tests has been performed with a population of 20 individuals. Selection is performed using the rank-based roulette operator (with the rank from 2.5 for the best individual to 1.6 for the worst individual - the step is 0.1). The crossover operator is one-point, with the probability 0.85. The elitist strategy with one elite individual is used in the generation replacement scheme. Multiple occurrences of an individual are removed from the population.

Two problem-specific *two-parts* mutation operator were used. The first operator *(TP1)* features two different probabilities of mutation for the two parts *(atoms, probabilities)* of an individual; after mutation, the real numbers in *probabilities* part of an individual have to be scaled since their sum must equal 1. The second operator *(TP2)* is a combination of ordinary mutation on *atoms* part, and a special mutation on *probabilities* part of an individual. Instead of performing mutation on two bits in the representation of *probabilities* part, two members p_{i1}, p_{i2} of *probabilities* part are chosen randomly and then replaced with random p'_{i_1}, p'_{i_2}, such that $p_{i_1} + p_{i_2} = p'_{i_1} + p'_{i_2}$ and $0 \leq p'_{i_1}, p'_{i_2} \leq 1$. The sum of probabilities does not change and no scaling is needed.

Four neighborhood structures are used within the VND-scheme [6,11]:

LS1 (LS denotes "local search"): For an individual M all the weight literals are divided into two sets: the first set (B) contains all satisfied literals, while the second one (W) contains all the remaining literals. The literal t_B ρ_B $c_B \in B$ (called the best one) with the biggest difference $|\mu(t_B) - c_B|$ between the left and the right side, and the literal t_W ρ_W $c_W \in W$ (the worst one) with the biggest difference $|\mu(t_W) - c_W|$ are found. Two sets of atoms are determined: the first set $B_{\text{At}(f)}$ contains all the atoms from M satisfying at least one classical formula α_i^B from $t_B = a_1^B w(\alpha_1^B) + \ldots + a_{k_B}^B w(\alpha_{k_B}^B)$, while the second one $W_{\text{At}(f)}$ contains all the atoms from M satisfying at least one classical formula α_i^W from $t_W = a_1^W w(\alpha_1^W) + \ldots + a_{k_W}^W w(\alpha_{k_W}^W)$. The probabilities of a randomly selected atom from $B_{\text{At}(f)} \setminus W_{\text{At}(f)}$ and a randomly selected atom from $W_{\text{At}(f)} \setminus B_{\text{At}(f)}$ are changed so that t_B ρ_B c_B remains satisfied, while the distance $|\mu(t_W) - c_W|$ is decreased or t_W ρ_W c_W is satisfied.

LS2: For na individual M, the *worst* weight literal t_W ρ_W c_W from W (the set of unsatisfied literals) with the biggest difference $|\mu(t_W) - c_W|$ is found. The literal can be represented as $\sum_{j=1}^{L} a_{Wj} \mu(at_j)$ ρ_W c_W. We try to change the vector of probabilities $[\mu(at_j)]$, so that the linear equation $\sum_{j=1}^{L} a_{Wj} \mu(at_j) = c_W$ is satisfied. The equation $\sum_{j=1}^{L} a_{Wj} \mu(at_j) = c_W$ represents a hyper-plane in R^n while $[a_{Wj}]$ denotes a vector normal to the hyper-plane. The projection of $[\mu(at_j)]$ to the hyper-plane - which satisfies the equation - is $[\mu'(at_j)] = [\mu(at_j)] + k_W [a_{Wj}]$. The calculation of k and the projection vector is simple and straight-forward $(k = \frac{c_w - a_W \circ [\mu(at_j)]}{|a_W|^2} = \frac{c_w - \sum_{j=1}^{L} \mu(at_j) a_{Wj}}{\sum_{j=1}^{L} a_{Wj}^2})$. We set the new vector of probabilities to be $[\mu''(at_j)] = \frac{[\max\{\mu'(at_j),0\}]}{\sum_{k=1}^{L} \max\{\mu'(at_k),0\}}$ (negative coordinates are replaced with 0, and the vector is scaled so that the sum of its coordinates $\sum_{j=1}^{L} \mu''(at_j)$ equals 1).

LS3 is similar to LS2, with the difference being made when choosing the weight literal t_W ρ_W c_W from W (the set of unsatisfied literals). The chosen literal is the one with the smallest difference $|\mu(t_W) - c_W|$; it is the *best bad literal*.

LS4 is similar to LS2 and LS3. Instead of calculating the projection $[\mu'(at_j)] = [\mu(at_j)] + k_W [a_{Wj}]$ for one chosen weight literal t_W ρ_W c_W from W, we calculate $k_{W_i} [a_{W_i j}]$ for each literal t_{W_i} ρ_{W_i} c_{W_i} from W (the set of unsatisfied literals) and calculate the *intermediate* vector $[\mu'(at_j)]$, by adding the linear combination to the original vector: $[\mu'(at_j)] = [\mu(at_j)] + \sum_{W_i} k_{W_i} [a_{W_i j}]$. The new vector of probabilities $[\mu''(at_j)]$ is then calculated in same fashion as in LS2.

4   Experimental Results

In this section we first present our test methodology and then we discuss the performed tests.

In our methodology, introduced in [13], the performance of the system is evaluated on a set of PSAT-instances, i.e., on a set of randomly generated formulas

Table 1. Results of tests with various mutation-types, mutation rates and pop. sizes

N,L, inst.no.	TP1 (12,4) 10 ind. no l.s.p.	TP2 (12,4) 10 ind. no l.s.p.	TP2 (24,8) 10 ind. no l.s.p.	TP2 (48,16) 10 ind. no l.s.p.	TP2 (12,4) 20 ind. no l.s.p.
50, 50, 1	5/52	5/89	5/61	5/126	5/35
50, 50, 2	5/9	5/14	5/7	5/3	5/3
50, 50, 3	5/220	5/249	5/456	5/231	5/61
50, 100, 1	5/73	5/100	5/104	5/167	5/47
50, 100, 2	5/286	5/262	5/442	5/602	5/144
50, 100, 3	5/996	5/643	5/1008	5/2626	5/261
50, 250, 1	5/707	5/592	5/979	5/1552	5/359
50, 250, 2	5/2486	5/1959	5/7222	0	5/1062
50, 250, 3	5/713	5/670	5/1292	5/2338	5/373
100, 100, 1	5/13	5/36	5/22	5/33	5/19
100, 100, 2	5/22	5/16	5/18	5/19	5/8
100, 100, 3	5/115	5/70	5/60	5/49	5/37
100, 200, 1	5/294	5/475	5/623	5/926	5/302
100, 200, 2	5/107	5/108	5/100	5/70	5/67
100, 200, 3	5/101	5/78	5/91	5/65	5/64
100, 500, 1	5/1470	5/1819	5/2882	5/4411	5/1131
100, 500, 2	5/3426	5/2881	5/4083	1/6470	5/1525
100, 500, 3	4/6817	5/4739	0	0	5/2667
200, 200, 1	0	4/2964	0	0	5/2428
200, 200, 2	5/117	5/278	5/162	5/208	5/159
200, 200, 3	5/12	5/126	5/143	5/124	5/81
200, 400, 1	5/20	5/166	5/103	5/82	5/80
200, 400, 2	3/4549	5/3428	0	0	5/2010
200, 400, 3	3/4377	5/3049	0	0	5/1733
200, 1000, 1	5/1403	5/2878	5/3619	5/4043	5/1912
200, 1000, 2	0	0	0	0	0
200, 1000, 3	5/1404	5/2863	5/3955	5/4384	5/1758

Table 2. Results of tests for pop.size = 10, TP2 mutation, with various mutation rates and different LS's

L, N, inst.no.	TP2 (12,4)				TP2 (24,8)				TP2 (48,16)			
	Local improvement procedure is applied in each generation											
	LS1	LS2	LS3	LS4	LS1	LS2	LS3	LS4	LS1	LS2	LS3	LS4
50, 50, 1	5/14	5/43	5/49	5/43	5/16	5/32	5/41	5/53	5/23	5/47	5/39	5/63
50, 50, 2	5/6	5/1	5/2	5/2	5/6	5/1	5/1	5/3	5/6	5/1	5/2	5/2
50, 50, 3	5/55	5/210	5/114	5/126	5/33	5/118	5/88	5/106	5/19	5/145	5/69	5/94
50, 100, 1	5/66	5/10	5/32	5/12	5/72	5/8	5/32	5/10	5/84	5/11	5/41	5/10
50, 100, 2	5/195	5/75	5/193	5/126	5/164	5/73	5/126	5/130	5/257	5/112	5/144	5/153
50, 100, 3	5/113	5/287	5/983	5/625	5/123	5/367	5/888	5/601	5/165	5/225	5/792	5/1119
50, 250, 1	5/709	5/415	5/291	5/524	5/975	5/341	5/209	5/460	5/1666	5/273	5/214	5/587
50, 250, 2	5/628	5/942	5/647	5/1604	5/681	5/1012	5/640	5/1738	5/2085	5/1165	5/1090	5/2170
50, 250, 3	5/444	5/232	5/280	5/230	5/447	5/143	5/240	5/209	5/394	5/155	5/227	5/150
100, 100, 1	5/22	5/1	5/2	5/5	5/25	5/1	5/1	5/5	5/22	5/1	5/1	5/4
100, 100, 2	5/18	5/11	5/7	5/8	5/20	5/5	5/9	5/7	5/18	5/3	5/7	5/5
100, 100, 3	5/71	5/47	5/49	5/49	5/57	5/37	5/55	5/53	5/68	5/31	5/41	5/44
100, 200, 1	5/354	5/107	5/283	5/182	5/258	5/115	5/187	5/173	5/208	5/60	5/205	5/194
100, 200, 2	5/31	5/130	5/86	5/100	5/35	5/114	5/88	5/44	5/34	5/51	5/27	5/65
100, 200, 3	5/99	5/30	5/64	5/82	5/96	5/17	5/19	5/25	5/69	5/16	5/53	5/46
100, 500, 1	5/1024	5/750	5/657	5/1458	5/626	5/784	5/614	5/1286	5/1046	5/556	5/341	5/2011
100, 500, 2	5/1598	5/1579	5/1320	5/1264	5/1548	5/1297	5/651	5/836	5/1063	5/1466	5/505	5/801
100, 500, 3	5/2088	5/3270	2/1763	5/2234	5/1462	5/3105	5/1938	5/2144	5/1546	5/3773	5/989	5/1730
200, 200, 1	3/2902	3/4447	4/2089	2/2829	5/1693	5/2268	5/2065	5/2691	5/1455	5/2076	5/1499	5/1206
200, 200, 2	5/395	5/271	5/181	5/273	5/293	5/185	5/125	5/178	5/217	5/118	5/84	5/125
200, 200, 3	5/157	5/26	5/60	5/41	5/144	5/17	5/39	5/13	5/155	5/12	5/37	5/24
200, 400, 1	5/41	5/58	5/40	5/153	5/30	5/26	5/50	5/62	5/34	5/32	5/13	5/38
200, 400, 2	0	5/2024	5/1562	5/2694	0	5/2063	5/1140	5/3686	0	5/3622	5/1045	0
200, 400, 3	0	5/1764	5/1090	5/1467	0	5/1309	5/692	5/1435	0	5/1412	5/631	5/1372
200, 1000, 1	5/3911	5/1160	5/1608	5/2545	5/3144	5/946	5/1183	5/2168	5/4167	5/630	5/511	5/2586
200, 1000, 2	0	0	0	0	0	0	0	0	0	0	1/2506	0
200, 1000, 3	5/4054	5/1658	5/897	3/2691	5/4016	5/1127	5/959	5/2285	4/4401	5/1254	5/1085	5/2115

in the wfc-form (with classical formulas in disjunctive normal form). The advantage of this approach is that a formula can be randomly generated according to the following parameters: N - the number of propositional letters, L - the number of weight literals, S - the maximal number of summands in weight terms, and D - the maximal number of disjuncts in DNF's of classical formulas. The considered set of test problems contains 27 satisfiable formulas. Three PSAT-instances were generated for each of 9 pairs of (N, L), where $N \in \{50, 100, 200\}$,

Table 3. Results of tests for pop.size = 10, TP2 mutation, with various mutation rates and LS's. LS's are used in each third generation cycle.

L, N, inst.no.	TP2 (12,4)				TP2 (24,8)				TP2 (48,16)			
	Local improvement procedure is applied in each third generation											
	LS1	LS2	LS3	LS4	LS1	LS2	LS3	LS4	LS1	LS2	LS3	LS4
50, 50, 1	5/32	5/40	5/63	5/54	5/21	5/37	5/52	5/58	5/23	5/39	5/41	5/28
50, 50, 2	5/7	5/3	5/3	5/3	5/7	5/3	5/4	5/4	5/4	5/2	5/4	5/3
50, 50, 3	5/63	5/141	5/109	5/138	5/56	5/193	5/105	5/151	5/48	5/186	5/63	5/87
50, 100, 1	5/99	5/18	5/36	5/10	5/92	5/19	5/33	5/12	5/97	5/25	5/37	5/20
50, 100, 2	5/148	5/92	5/144	5/126	5/173	5/114	5/193	5/98	5/232	5/215	5/115	5/90
50, 100, 3	5/184	5/325	5/446	5/461	5/181	5/237	5/607	5/477	5/330	5/560	5/1551	5/1435
50, 250, 1	5/749	5/457	5/279	5/478	5/759	5/420	5/282	5/525	5/1677	5/626	5/312	5/464
50, 250, 2	5/863	5/1029	5/763	5/1302	5/935	5/763	5/803	5/2215	5/1510	5/3307	5/1937	4/5124
50, 250, 3	5/327	5/268	5/257	5/249	5/297	5/205	5/261	5/207	5/923	5/173	5/173	5/151
100, 100, 1	5/25	5/3	5/3	5/6	5/27	5/3	5/5	5/10	5/25	5/3	5/4	5/9
100, 100, 2	5/12	5/5	5/13	5/11	5/18	5/8	5/13	5/11	5/12	5/11	5/11	5/10
100, 100, 3	5/88	5/47	5/63	5/58	5/56	5/41	5/43	5/58	5/50	5/37	5/31	5/37
100, 200, 1	5/271	5/131	5/295	5/210	5/278	5/111	5/181	5/156	5/234	5/94	5/296	5/137
100, 200, 2	5/45	5/146	4/42	5/142	5/38	5/67	5/48	5/97	5/36	5/44	5/24	5/41
100, 200, 3	5/90	5/46	5/51	5/64	5/71	5/25	5/74	5/58	5/51	5/35	5/29	5/31
100, 500, 1	5/787	5/1147	5/883	5/1286	5/766	5/821	5/464	5/1377	5/1028	5/895	5/306	5/1714
100, 500, 2	5/1393	5/2565	5/1061	5/1409	5/1386	5/1436	5/856	5/846	5/1559	5/1601	5/517	5/730
100, 500, 3	5/2375	5/3116	2/1649	5/2598	5/1327	5/3429	5/1452	5/2130	5/2199	5/4200	5/734	5/2029
200, 200, 1	3/1685	4/3877	4/3073	5/4482	5/2706	5/1567	4/3018	4/1722	5/1959	5/2161	5/2064	4/1917
200, 200, 2	5/520	5/323	5/384	5/285	5/301	5/226	5/146	5/278	5/220	5/118	5/97	5/114
200, 200, 3	5/186	5/25	5/49	5/40	5/154	5/35	5/39	5/54	5/88	5/26	5/29	5/34
200, 400, 1	5/74	5/64	5/64	5/145	5/40	5/50	5/46	5/64	5/51	5/36	5/16	5/53
200, 400, 2	0	5/2226	5/1950	5/2922	0	5/2189	5/1260	5/3915	0	4/4194	5/2138	0
200, 400, 3	0	5/1954	5/1549	5/1912	0	5/1738	5/1099	5/1345	0	5/1801	5/864	5/1584
200, 1000, 1	5/3783	5/1353	5/1505	5/2458	5/3643	5/921	5/1040	5/2225	5/4104	5/1155	5/675	5/2763
200, 1000, 2	0	0	0	0	0	0	2/4313	0	0	0	1/3923	0
200, 1000, 3	5/3880	5/1754	5/1520	5/2397	5/3774	5/1455	5/1011	5/2361	4/4652	5/1552	5/496	5/2519

Table 4. Results of tests for pop.size = 10, TP2 mutation, and LS's. LS's are used after each generation cycle. No mutation of *probabilities* part - it is reinitialized after each five generation cycle.

L, N, inst.no.	TP2 (12,0)				TP2 (24,0)				TP2 (48,0)			
	LS is applied in each generation. *Probabilities* reinitialized in each fifth generation.											
	LS1	LS2	LS3	LS4	LS1	LS2	LS3	LS4	LS1	LS2	LS3	LS4
50, 50, 1	5/64	5/78	5/114	5/100	5/23	5/56	5/90	5/122	5/21	5/40	5/68	5/135
50, 50, 2	5/10	5/1	5/1	5/3	5/7	5/1	5/2	5/2	5/5	5/1	5/1	5/3
50, 50, 3	5/72	5/2652	5/515	5/905	5/87	5/592	5/719	5/1649	5/34	5/596	5/387	5/1317
50, 100, 1	5/244	5/11	5/66	5/6	5/144	5/7	5/52	5/8	5/149	5/10	5/51	5/8
50, 100, 2	5/828	5/493	5/987	5/942	5/626	5/248	5/589	5/534	5/417	5/165	5/380	5/352
50, 100, 3	5/1134	4/3436	2/3807	2/3231	5/399	5/3016	5/4837	5/2809	5/230	5/3001	2/8483	5/4084
50, 250, 1	5/3533	5/1967	5/1663	5/1709	5/1776	5/1247	5/1512	5/1389	5/1864	5/1412	5/966	5/824
50, 250, 2	5/4940	5/5186	4/5444	5/4309	5/5155	5/4750	5/4088	4/3380	5/3908	3/3440	4/2608	5/2904
50, 250, 3	5/2747	5/1475	5/1379	5/1821	5/2098	5/808	5/961	5/885	5/1755	5/648	5/701	5/584
100, 100, 1	5/17	5/1	5/2	5/5	5/25	5/1	5/1	5/8	5/22	5/1	5/1	5/6
100, 100, 2	5/29	5/7	5/7	5/8	5/16	5/6	5/19	5/11	5/16	5/7	5/15	5/15
100, 100, 3	5/182	5/130	5/216	5/136	5/161	5/84	5/139	5/119	5/107	5/97	5/114	5/117
100, 200, 1	5/4340	5/175	5/1341	5/669	5/3398	5/206	5/756	5/556	5/1757	5/146	5/999	5/370
100, 200, 2	5/44	5/193	5/659	5/343	5/24	5/164	5/273	5/139	5/44	5/98	5/101	5/118
100, 200, 3	5/272	5/115	5/82	5/93	5/217	5/74	5/124	5/63	5/221	5/57	5/55	5/80
100, 500, 1	1/6180	5/4608	2/4928	0	3/3862	5/2889	5/3835	5/6362	5/2450	5/1561	5/2541	5/4918
100, 500, 2	0	2/6595	0	1/5478	1/6055	4/2248	3/4932	4/5190	1/5834	4/3007	5/4221	5/2323
100, 500, 3	0	0	0	0	0	0	0	0	2/5650	0	2/3946	0
200, 200, 1	0	0	0	0	0	0	0	0	0	0	0	0
200, 200, 2	4/1066	5/2030	5/983	4/2421	4/1859	5/723	5/1244	4/2807	5/934	5/680	5/655	5/1367
200, 200, 3	5/814	5/199	5/179	5/292	5/382	5/159	5/201	5/284	5/294	5/47	5/109	5/218
200, 400, 1	5/96	5/122	5/343	5/334	5/110	5/183	5/323	5/262	5/70	5/172	5/247	5/134
200, 400, 2	0	0	0	0	0	0	0	0	2/3155	2/4084	0	0
200, 400, 3	0	0	0	0	1/1602	0	0	0	1/1673	0	1/3333	0
200, 1000, 1	0	0	0	0	1/4398	2/4147	0	0	2/3025	5/3439	2/3660	0
200, 1000, 2	0	0	0	0	0	0	0	0	0	0	0	0
200, 1000, 3	0	0	0	0	0	0	0	1/3328	2/2083	4/4012	1/3909	1/4709

and $L \in \{N, 2N, 5N\}$. For every instance $S = D = 5$. Having the above parameters, L atoms and their probabilities (with the constraint that the sum of probabilities must be equal to 1) are chosen. Next, a formula f containing L basic weight formulas is generated. It contains primitive propositions from the set $\{p_1, \ldots, p_N\}$ only. Every weight literal contains at most S summands in its

Table 5. Results of tests for pop.size = 10, TP2 mutation, and VND. In each generation, LS's are used in cycle, as long as any of them makes progress.

L, N, inst.no.	TP2(12,4)	TP2(24,8)	TP2(48, 16)
	Combination of all LS's is applied in each generation.		
50, 50, 1	5/11	5/6	5/7
50, 50, 2	5/1	5/1	5/1
50, 50, 3	5/11	5/30	5/15
50, 100, 1	5/5	5/5	5/8
50, 100, 2	5/109	5/73	5/76
50, 100, 3	5/68	5/116	5/121
50, 250, 1	5/344	5/249	5/235
50, 250, 2	5/552	5/393	5/647
50, 250, 3	5/89	5/77	5/72
100, 100, 1	5/1	5/1	5/1
100, 100, 2	5/6	5/4	5/5
100, 100, 3	5/39	5/37	5/28
100, 200, 1	5/37	5/34	5/37
100, 200, 2	5/64	5/17	5/23
100, 200, 3	5/29	5/18	5/17
100, 500, 1	5/415	5/330	5/221
100, 500, 2	5/370	5/290	5/297
100, 500, 3	5/1473	5/829	5/555
200, 200, 1	2/2209	5/3036	4/1618
200, 200, 2	5/212	5/134	5/92
200, 200, 3	5/60	5/26	5/32
200, 400, 1	5/22	5/28	5/15
200, 400, 2	5/1642	5/1977	5/2447
200, 400, 3	5/1088	5/604	5/479
200, 1000, 1	5/965	5/728	5/635
200, 1000, 2	0	1/4537	4/3362
200, 1000, 3	5/1197	5/1065	5/1014

Table 6. Average time (rounded to seconds) used by the test computer to execute *succesful tests for some selected parameters.* (Npte: Value 0 means that the average time was less than half second.)

L, N, inst. no.	Table 1 TP2(12,4) No LS 10 ind.	20 ind.	Table 2 TP2(12,4) 10 individuals LS's applied in each generation LS1	LS2	LS3	LS4	Table 3 TP2(12,4) 10 individuals LS's applied in each third generation LS1	LS2	LS3	LS4	Table 5 TP2(12,4)	TP2(24,8)	TP2(48,16) 10 individuals Combination of LS's applied in each generation
50, 50, 1	0	1	0	0	0	0	0	0	0	0	0	0	0
50, 50, 2	0	1	0	0	0	0	0	0	0	0	0	0	0
50, 50, 3	0	1	0	1	1	0	0	0	0	0	0	0	0
50, 100, 1	1	1	1	0	0	2	1	0	0	1	1	0	1
50, 100, 2	1	2	1	1	2	2	1	1	1	2	2	2	3
50, 100, 3	3	3	1	2	7	10	1	2	3	4	1	3	3
50, 250, 1	16	20	28	16	16	39	22	14	11	21	40	35	42
50, 250, 2	51	56	24	38	34	97	26	35	30	50	68	70	132
50, 250, 3	18	20	18	9	17	25	10	8	13	14	15	16	19
100, 100, 1	0	1	0	0	0	0	0	0	0	0	0	0	0
100, 100, 2	0	1	0	0	0	0	0	0	0	0	0	0	0
100, 100, 3	0	1	0	0	0	1	1	0	0	0	1	1	1
100, 200, 1	8	12	10	3	8	9	6	3	8	7	5	5	7
100, 200, 2	2	3	1	3	2	4	1	3	1	4	4	1	2
100, 200, 3	1	3	4	1	2	26	2	1	1	2	2	2	2
100, 500, 1	187	236	170	130	149	384	94	145	244	228	269	294	271
100, 500, 2	295	309	242	241	298	333	169	306	151	228	236	260	480
100, 500, 3	484	575	326	509	416	775	296	390	355	461	1019	777	671
200, 200, 1	58	91	71	108	56	134	34	78	66	3471	146	270	202
200, 200, 2	5	6	11	7	7	14	11	7	10	9	13	11	9
200, 200, 3	2	3	4	1	2	2	4	1	1	4	4	2	3
200, 400, 1	12	11	4	6	5	25	6	7	5	14	8	11	7
200, 400, 2	238	286	N/A	195	163	484	N/A	171	161	296	479	686	1128
200, 400, 3	205	230	N/A	174	205	247	N/A	153	201	208	419	334	374
200, 1000, 1	1593	2173	3064	888	1347	2972	2307	811	1271	1865	2363	2087	2032
200, 1000, 2	N/A	N/A	N/A	N/A	N/A	N/A	N/A	N/A	N/A	N/A	N/A	19582	19977
200, 1000, 3	1489	1861	3298	1364	792	3548	2456	1135	1080	2023	2818	2770	2778

weight term. Every classical formula is in disjunctive normal form with at most D disjuncts, while every disjunct is a conjunction of at most N literals. For every weight term t coefficients are chosen, and the value of t is computed. Next, the sum $sp(t)$ of positive coefficients and the sum $sn(t)$ of negative coefficients are computed. Finally, the right side value of the weight literals between $sp(t)$ and $sn(t)$, and the relation sign are chosen such that f is satisfiable.

We have implemented some versions of GA-VND-PSAT. The variants differ in the type of the mutation operator, the mutation probability, and the neighborhood structure used within the VND-scheme. All results are averages over 5 independent trials for each problem instance.

All the tests were run on the Pentium P4(2.4GHz,512MB)-based Linux station. We decided that it would be better to test more problem instances of different sizes (even very large scale instances) rather than making more trials on a smaller set of instances (of smaller or average size). Since the tests are of large size, the necessity to perform them in a reasonable time imposed to set the maximal number of generations to be: 10000 for $N = 50$, 7000 for $N = 100$ and 5000 for $N = 200$.

The results are summarized in the tables 1 - 5. Each table entry contains the number of successful trials (out of 5) and the average number of generations in successful trials. The first column in all tables contains the problem instance data. TP1 and TP2, followed by two numbers, denote the type and the rate of two-part mutation (see Section 3). The two numbers are the expected average (per generation/individual) numbers of bit mutations in atoms and probabilities parts, respectively.

Table 1 contains results of tests with no LS involved. Three central columns show results for TP2 mutation, population of 10 individuals and three different mutation rates. The second column on the left contains the results of tests with TP1 mutation used instead of TP2. The rightmost column contains the results with population of 20 individuals.

Comparison of columns 2 and 3 shows that, while with TP1 mutation the solution was found in less generations (on average) than with TP2, on some instances TP2 outperformed TP1. Comparison of columns 3 and 6 shows that doubling the size of population did not increase the performance by same factor.

Due to these observations, we chose to use TP2 mutation and population of 10 individuals. The three central columns of Table 1 show the result for those parameters, with three different mutation rates. For most instances, increasing mutation rates did not give the expected results and has even deteriorated the performance of GA.

Table 2 shows the results of applications of LS's. A chosen LS is applied to each individual, in each generation, and is repeated as long as it makes improvement, i.e. until the $d(M)$ for the individual decrease. If no improvement is made, altered temporary individual is discarded, and the original is kept.

Table 2 can be compared to columns 3-6 of Table 1. Results show that - on average as well as for 'harder' instances - LS3 gives the best results. That can be explained by the fact that LS3 makes the smallest changes to an individual; it tries to make an individual to satisfy the best bad literal. Note that increased mutation rate increases average success rate and performance (in contrast with GA with no LS). The 'hardest' instance (200, 1000, 2) was solved in one test, using

LS3. Also note that GA with LS1 performed much worse - for some instances - than the pure GA.

The application of LS's increases the computational time needed for one generation cycle. We tried to decrease computational time, while keeping the benefits of LS's, by decreasing the use of LS's. For the next set of tests, reported in Table 3, we used an LS in each third generation cycle, instead of in each generation cycle. The average number of generations (needed to find a solution) was not drastically increased (compare 179 to 135), while the success rate increased (1514 successful tests compared to 1511). Instance *(200, 1000, 2)* was solved in three tests, again using LS3.

The next set of tests was designed to analyze the contribution of the mutation of the *probabilities part* to the algorithm. We eliminated that mutation by setting the second parameter of TP2 mutation to 0. The *probabilities part* is reinitialized (i.e. new random values are set in the same way as during the initialization of GA) after each five generation cycle, to avoid local minima. The results are shown in Table 4. Poor performance clearly indicates the importance of *probabilities part* mutation. Different length of reinitialization period might give better results; since this set of tests was much more time consuming than other tests, we decided to leave that research direction for the future.

LS2 had the best average performance, when used alone. However, for some test instances, another local searches show better performances. That is the reason why the last set of tests involve a combination of all LS's within the VND scheme: in each generation, LS1 is applied to an individual (repeatedly, while it makes improvement), followed by LS2, LS3 and LS4. LS's are reapplied in a cycle. The cycle is stopped when four consecutive LS's make no improvement, i.e. the particular individual can not be improved any more (using any of the four LS's). Results are shown in Table 5. Although this combination of LS's increases evaluation costs (per a generation cycle), it significantly increases the success rates and decreases the number of generation cycles needed to find the solution.

The last table 6 contains the average running time of successful tests as measured on our test computer. Table shows running times only for selected tests, due to space limitations. Columns 2 and 3 show times for tests without LS's, with different population size (10 individuals vs 20 individuals). Increased population size does result in smaller number of iterations needed to find the solution, but the computational cost for each iteration is increased and the overall computational cost is greater than with smaller population size. In columns 4-7 and 8-11 we can compare the efficiency of various LS's. It is clear that LS2 and LS3 are more efficient than LS1 and LS4 when used for large problem instances, however it is not clear which of them is the most efficient. The running times in columns 8-11 (LS's applied in each third generation) are on average smaller than times in columns 4-7 (LS's applied in each generation). However, this does not mean that the principle of reducing application of LS's to each third generation is always more efficient. Finally, columns 12-14 show execution times for tests using combination of LS's. Combined usage of LS's is not justified in terms of time efficiency, but it is justified in terms of increased success rate. Higher mutation rate in this GA-VND setup leads to better time efficiency and higher success rate, except for a few less complex problem instances.

5   Conclusion

In this paper we have described and tested a hybrid GA and VND-based satisfiability checker GA-VND-PSAT for probabilistic logic trying to make a trade-off between completeness and computation time. We have performed some experimental analysis and tested GA-VND-PSAT against GA-PSAT. The results of our tests show both the higher success rate and the computational superiority of GA-VND-PSAT. Although a careful implementation contributes to better properties of GA-VND-PSAT, the fact that GA-VND-PSAT outperforms GA-PSAT is in the first place due to the new PSAT-specific operators and heuristics. GA-VND-PSAT is able to solve PSAT-instances with the number of propositional letters $N = 200$ and the number of weight literals $L = 1000$. We are not aware of any larger PSAT-instances reported in the literature. For example, $N = 50$, $L = 70$ in [9], $N = 140$, $L = 300$ in [8], L is up to 500 in [5], and $N = 200$, $L = 800$ in [7]. Also, we note that the instances considered in the mentioned papers have simpler form than ours, since there S, the maximal number of summands in weight terms, and D, the maximal number of disjuncts in DNF's of classical formulas, are set to be 1 and 4 (or 3), respectively (we have $S = D = 5$), while the used classical formulas are clauses (i.e., disjunctions of propositional letters and their negations - propositional literals). In other words, their weight term contains the probability of only one clause with up to 4 propositional literals. Nevertheless, GA-PSAT and GA-VND-PSAT are only the first steps in our investigations, and there are many directions for further research. The first one obviously concerns improving the problem specific GA-operators, and heuristics used within the VND-schema. The next problem could be to see how our GA-VND-approach performs on the so-called interval PSAT [5] in which weight terms belong to intervals of probability, i.e., basic weight formulas are of the form $c_1 \geq t \geq c_2$. Another interesting issue could be to consider how our GA-VND-approach to PSAT can be extended for a more expressible version of probabilistic logic which allows iteration of probabilistic operators [3,12] in which case PSAT is PSPACE-complete, or for the framework of conditional probabilities. Finally, our test-generating methods can be seen as a generalization of the standard SAT-generating approach, but we have more parameters than in classical propositional logic. Our experiences show that increasing L leads to harder problems, since L is the number of constraints. However, still we are not able to conjecture any relation between N, L and the other parameters and the hardness of problems (for example, similar to the phase transition phenomenon for SAT). Thus, it is clear that far more tests with a much greater number of problems and an exhaustive study should be done.

References

1. Evolutionary computation, Vol. I, II, T. Baeck, D. B. Fogel, and T. Michalewicz edts., Institute of Physics Publishing, Bristol and Philadelphia, 2000.
2. R. Fagin, J. Halpern, and N. Megiddo. A logic for reasoning about probabilities.Information and Computation, 87:78–128, 1990.

3. R. Fagin and J. Halpern. Reasoning about knowledge and probability. Journal of the ACM, 41(2), 340–367, 1994.
4. G. Georgakopoulos, D. Kavvadias, and C. Papadimitriou. Probabilistic satisfiability. Journal of Complexity, 4(1):1–11, 1988.
5. P. Hansen, B. Jaumard, Probabilistic satisfiability. in Gabbay, Dov M. (ed.) et al., Handbook of defeasible reasoning and uncertainty management systems. Vol. 5: Algorithms for uncertainty and defeasible reasoning. Dordrecht: Kluwer Academic Publishers. 321–367, 2000.
6. P. Hansen, N. Mladenović, Variable neighborhood search, Fred in W. Glover and Gary A. Kochenberger (edts.), Handbook of Metaheuristics,International Series in Operations Research and Management Science, Vol. 57, Springer, 145–184, 2003.
7. P. Hansen, S. Perron. Merging the Local and Global Approaches to Probabilistic Satisfiability, GERAD tech. report, 2004.
8. B. Jaumard, P. Hansen, and M. P. de Aragao. Column generation methods for probabilistic logic. ORSA Journal on Computing, 3:135–147, 1991.
9. D. Kavvadias, and C. Papadimitriou. A linear programming approach to reasoning about probabilities. Annals of Mathematics and Artificial Intelligence, vol. 1, No.1-4, 189–205, 1990.
10. N. Nilsson. Probabilistic logic. Artificial Intelligence, 28:71-87, 1986.
11. N. Mladenović, P. Hansen, Variable neighborhood search, Computers and Operations Research, vol. 24, no. 11, 1097–1100, 1997.
12. Z. Ognjanović and M. Rašković. Some first-order probability logics. Theoretical Computer Science, 247(1-2):191-212, 2000.
13. Z. Ognjanović, J. Kratica, M. Milovanović. A genetic algorithm for satisfiability problem in a probabilistic logic: A first report. LNCS 2143, 805 - 816, 2001.
14. Z. Ognjanović, U. Midić, J. Kratica. A genetic algorithm for probabilistic SAT problem. LNCS 3070, 462–467, 2004.

A Hybrid Meta-heuristic Approach for Natural Gas Pipeline Network Optimization

C. Borraz-Sánchez and R.Z. Ríos-Mercado

Universidad Autónoma de Nuevo León, Graduate Program in Systems Engineering, AP 111 – F,
Cd. Universitaria, San Nicolás de los Garza, NL 66450, México
{conrado,roger}@yalma.fime.uanl.mx

Abstract. In this paper we propose a hybrid heuristic solution procedure for fuel cost minimization on gas transmission systems with a cyclic network topology, that is, networks with at least one cycle containing two or more compressor station arcs. Our heuristic solution methodology is based on a two-stage iterative procedure. In a particular iteration, at a first stage, gas flow variables are fixed in each network arc and optimal pressure variables in each network node are found via non-sequential dynamic programming. At a second stage, pressure variables are fixed and a short-term memory Tabu Search procedure is used for guiding the search in the flow variable space. Empirical evidence supports the effectiviness of the proposed procedure outperforming the best existing approach to the best of our knowledge.

Keywords: steady state, natural gas, transmission networks, non-convex problem, dynamic programming, tabu search.

1 Introduction

In this paper, we address the problem of minimizing the fuel consumption incurred by compressor stations in a natural gas pipeline transmission system. During this process, energy and pressure are lost due to both friction between the gas and the pipes' inner wall, and heat transfer between the gas and the environment. To keep the gas flowing through the system, it is necessary to periodically increase its pressure, so compressor stations are installed through the network. It is estimated that compressor stations typically consume about 3 to 5% of the transported gas. This transportation cost is significant because the amount of gas being transported in large-scale systems is huge. In the other hand, even a marginal improvement in gas operations can have a significant positive impact from the economic standpoint, so this provides the main motivation from the practical side for the proposed work.

This problem is represented by a network, where arcs correspond to pipelines and compressor stations, and nodes correspond to their physical interconnection points. We consider two types of continuous decision variables: mass flow rates through each arc, and gas pressure level at each node. So, from the optimization perspective, this problem is modeled as a nonlinear program (NLP), where the cost function is typically nonlinear and non-convex, and the set of constraints is typically non-convex as well. It is well

M.J. Blesa et al. (Eds.): HM 2005, LNCS 3636, pp. 54–65, 2005.

know that non-convex NLP is NP-hard [6]. This motivates the choice of the proposed heuristic approach.

The state of the art on research on this problem reveals a few important facts. First, there are two fundamental types of network topologies: non-cyclic and cyclic. We would like to emphasize that, the former is a type of topology that has received most of the attention during the past 30 years. Several methods of solution have been developed, most of them based on Dynamic Programming (DP), which were focused on non-cyclic networks.

In particular, as far as handling cyclic topologies is concerned, gradient search and DP approaches have been applied with little or limited success. The main limitation of the former is its local optimality status. The drawback of the latter, is that its application is limited to problems where the flow variables are fixed, so the final solution is "optimal" with respect to a pre-specified set of flow variables. This is because cyclic topologies are a lot harder to solve.

In this paper, we proposed a novel solution methodology for addressing the problem of how to optimally operate the compressor stations in a natural gas pipeline system, focusing in cyclic topologies. The proposed technique combines a non-sequential DP technique (originally proposed by Carter [2]) within a Tabu Search (TS) framework. For the past twelve years, TS has established its position as an effective meta-heuristic guiding the design and implementation of algorithms for the solution of combinatorial optimization problems in a number of different areas (Glover and Laguna [5]). A key reason for this success is the fact that the algorithm is sufficiently flexible to allow designers to exploit prior domain knowledge in the selection of parameters and sub-algorithms. In this case, even though we are dealing with a continuous optimization problem, the high non-convexity of the objective function and the versatility of TS to overcome local optimality make TS very attractive with an appropriate discrete solution space.

Empirical evidence over a wide range of instances with data taken from industry shows the efficiency of the proposed approach. A comparison with former approaches which include GRG-based and state-of-the-art Carter's DP technique demonstrates the significant superiority of our procedure. Furthermore, in order to assess the quality of the solutions delivered by our procedure, a lower bound procedure was derived. It is shown that the optimality gaps found by our technique are less than 16%, most of them less than 10%, which represents a significant progress to the current state of the art in this area. The scientific contribution of this work is providing the best technique known to date, to the best of our knowledge, for addressing this type of problem in cyclic topologies.

The rest of this paper is organized as follows. In Section 2, we formally introduce the fuel consumption minimization problem (FCMP), describing its main features, modeling assumptions, and important properties. Then, in Section 3, we present a review of earlier approaches for this problem, highlighting the most related to our work, and how we attempt to exploit some of them. The proposed methodology is fully described in Section 4. An extensive computational evaluation of the heuristic, including comparison with earlier approaches, is presented in Section 5. Finally, we wrap up this work with the conclusions and directions for future research in Section 6.

2 Problem Description

Pipeline system models can be mainly classified into steady-state and transient systems. The difference between the two is as follows. The flow dynamics through a pipeline is ruled by a partial differential equation involving derivatives with respect to time. Under a steady-state assumption, it is possible to work out this equation and reduce to a nonlinear equation with no derivatives, which makes the problem a lot more tractable from the optimization perspective. Like all those previous works (reviewed in Section 3), here we assume a steady-state model. That is, our model provides solutions for systems that have been operating for a relatively large amount of time, which is a common practice in industry. Transient analysis has been done basically by descriptive models, so optimization for transient systems remains as one of the great research challenges in this area. We also assume we work with a deterministic model, that is, each parameter is known with certainty, which is a very reasonable assumption. In terms of the compressor stations, we assume we work with centrifugal compressor units, which are the most commonly found in industry. As far as the network model is concerned, we assumed the network is balanced, that is, no gas is lost, and that each arc in the network has a pre-specified direction.

2.1 The Model

This model was originally introduced by Wu et al. [19].

Sets
   V: Set of all nodes in the network
   V_s: Set of supply nodes ($V_s \subset V$)
   V_d: Set of demand nodes ($V_d \subset V$)
   A_p: Set of pipeline arcs
   A_c: Set of compressor station arcs
   A: Set of all arcs in the network; $A = A_p \cup A_c$

Parameters
   U_{ij}: Arc capacity of pipeline (i,j); $(i,j) \in A_p$
   R_{ij}: Resistance of pipeline (i,j); $(i,j) \in A_p$
   P_i^L, P_i^U: Pressure lower and upper limits at each node; $i \in V$
   B_i: Net mass flow rate at node i; $i \in V$. $B_i > 0$ if $i \in V_s$, $B_i < 0$ if $i \in V_d$,
       $B_i = 0$ otherwise

Variables
   x_{ij}: Mass flow rate in arc (i,j); $(i,j) \in A$
   p_i: Pressure at node i; $i \in V$

Formulation

   (FCMP)

   Minimize $\displaystyle\sum_{(i,j)\in A_c} g_{ij}(x_{ij}, p_i, p_j)$ (1)

   subject to $\displaystyle\sum_{j:(i,j)\in A} x_{ij} - \sum_{j:(j,i)\in A} x_{ji} = B_i$ $i \in V$ (2)

$$x_{ij} \leq U_{ij} \qquad (i,j) \in A_p \qquad (3)$$

$$p_i^2 - p_j^2 = R_{ij} x_{ij}^2 \qquad (i,j) \in A_p \qquad (4)$$

$$p_i \in [p_i^L, p_i^U] \quad i \in V \qquad (5)$$

$$(x_{ij}, p_i, p_j) \in D_{ij} \qquad (i,j) \in A_c \qquad (6)$$

$$x_{ij}, p_i \geq 0 \qquad (i,j) \in A, \, i \in V \qquad (7)$$

The objective function (1) represents the total amount of fuel consumption in the system. Constraints (2)-(3) are the typical network flow constraints representing node mass balance and arc capacity, respectively. Constraint (4) represents the gas flow dynamics in each pipeline under the steady-state assumption. Constraints (5) denote the pressure limits in each node. These limits are defined by the compressor physical properties. Constraint (6) represents the non-convex feasible operating domain D_{ij} for compressor station (i,j). Finally, the mathematical model is bounded by non-negative decision variables (7). The algebraic representation of D_{ij} is the result of curve fitting methods based on empirical data taken from the compressors.

For measuring fuel consumption, we use a function g_{ij} in the following form:

$$g(x_{ij}, p_i, p_j) = \alpha x_{ij} \left\{ \left(\frac{p_j}{p_i} \right)^m - 1 \right\}, \quad (x_{ij}, p_i, p_j) \in D_{ij},$$

where α and m are assumed constant (and known) parameters that depend on the gas physical properties. A more detailed study on the nature of both the compressor station domain and the fuel consumption function is given in [19].

3 Previous Work

In this section, we review the most significant contributions over the last 30 years for solving the FCMP.

3.1 Methods Based on Dynamic Programming

The key advantages of DP are that a global optimum is guaranteed to be found and that nonlinearity can be easily handled. In contrast, its application is practically limited to non-cyclic networks, such as linear (also known as gun-barrel) or tree topologies, and that computation increases exponentially in the dimension of the problem, commonly referred as the curse of dimensionality.

DP for pipeline optimization was originally applied to gun-barrel systems in the late 1960s. It has been one of the most useful techniques due to both its computational behavior and its versatility for handling non-linearity on sequential systems. DP was first applied to linear systems by Wong and Larson [16] in 1968, and then applied to tree-structured topologies by Wong and Larson [17]. A similar approach was described by Lall and Percell [7] in 1990, who allow one diverging branch in their system.

The most significant work on cyclic networks know to date is due to Carter [2] who developed a non-sequential DP algorithm, but limited to a fixed set of flows. In our work, we use Carter's ideas and incorporate them within a Tabu Search scheme for iteratively adjusting the set of flows with great success. This will be further described in Section 4.

3.2 Methods Based on Gradient Search

In 1987, Percell and Ryan [11] applied a different methodology based on a Generalized Reduced Gradient (GRG) non-linear optimization technique for non-cyclic structures. One of the advantages of GRG, when compared with DP, is that they can handle the dimensionality issue relatively well, and thus, can be applied to cyclic structures. Nevertheless, being a method based on a gradient search, there is no guarantee for a global optimal soultion, especially when there are discrete decision variables. Villalobos-Morales and Ríos-Mercado [15] evaluated preprocessing techniques for GRG, such as scaling, variable bounding, and choice of starting solution, that resulted in better results for both cyclic and non-cyclic structures. More recently, Flores-Villarreal and Ríos-Mercado [4] performed an extensive computational evaluation of the GRG method over a large set of instances on cyclic structures with relative success. No comparison to DP was done in that work, so part of our contribution is to provide a comparison frame among Carter's NDP, GRG, and our method tested in the same set of instances.

3.3 Other Approaches

Wu, Boyd, and Scott [18] presented a mathematical model for the fuel cost minimization on a compressor station with a single unit. It was the first work that fully addressed the mathematical description of a centrifugal compressor. Later, Wu et al. [19] completed the analysis for the same problem, but considering several units within compressor stations. In a related work, some of the most important theoretical properties regarding pipeline networks are developed by Ríos-Mercado et al. [13].

In a variation of this problem, Cobos-Zaleta and Ríos-Mercado [3] recently presented a solution technique based on an outer approximation with equality relaxation and augmented penalty algorithm OA/ER/AP for solving a mixed-integer non-linear programming model, where an integer decision variable, representing the number of compressor units running within each station, is incorporated. They present satisfactory results as they were able to find local optima for many instances tested.

Optimization techniques have also been applied for transient (time dependent) models (e.g., Osiadacz [8], and Osiadacz and Swierczewski [10]), and network design (e.g., Osiadacz and Górecki [9]), with modest success. See Ríos-Mercado [12] for more references on optimization techniques applied to gas pipeline problems. It is important to mention that optimization approaches developed to date work well under some general assumptions; however, as the problems become more complex, the need arises for further research and effective development of algorithms from the optimization perspective.

4 Solution Procedure

Basically, the proposed methodology consists of four components: (a) Preprocessing: This phase is performed both to refine the feasible operating domain given by tightening decision variable bounds, and to reduce the size of the network by a reduction technique (motivated by the work of Ríos-Mercado et al. [13]); (b) Finding an initial feasible flow: In this phase, a set of feasible flows is found by two different methods: a classic

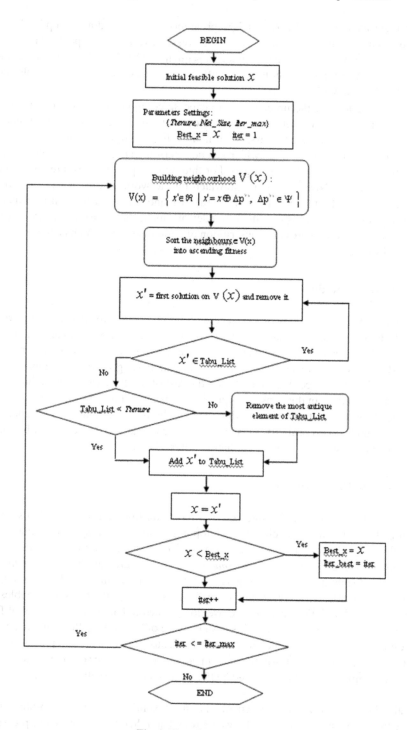

Fig. 1. Flowchart of NDPTS

assignment technique and a reduced graph algorithm; (c) Finding an optimal set of pressure values: In this phase, a set of optimal pressures (for the pre-specified flow in the previous phase) is found by applying a non-sequential DP (NDP) algorithm; (d) Flow modification: Here, an attempt to find a different set of flows is made by employing a tabu search framework.

So the key idea of the procedure is to execute components (c) and (d) iteratively until a stopping criteria is satisfied. As we know from theoretical properties of pipeline networks [13], step (d) is unnecessary for non-cyclic topologies because there exists a unique set of optimal flow values which can be determined in advance at preprocessing. So, here we focus on cyclic topologies. For finding the optimal set of pressures in (c), we implemented a NDP technique motivated by the work of Carter [2]. The overall procedure is called NDPTS. Components (a), (b), and (c) are fairly well documented in our previous work [1], so, in the reminder of this section, we assume we have an initial feasible flow and provide a description of component (d), which is the core of the proposed work.

4.1   Overall Procedure

Figure 1 shows a flow chart of the general solution procedure. Briefly, we start the procedure by finding an initial feasible set of flows x by the NDP algorithm. Then a list of neighbors of x, $V(x)$, is generated. To build $V(x)$ we take a mass flow rate in a selected arc belonging to a cycle and modify it by increasing or decreasing its value by Δ_x units. Note that once this value its set, the rest of the flow variables in the arc are easily determined, so in this sense, it is precisely this mass flow rate which becomes the attribute. Then the best $x' \in V(x)$ which is not tabu is chosen and the corresponding subsets are updated accordingly. This process of local search and selection of best non-tabu neighbor is repeated until a termination criteria is met.

4.2   Proposed Tabu Search

We define the nature of a feasible solution based on three basic components which are directly related with a cyclic network topology: (a) *static component*, a mass flow rate value not belonging to any cycle; (b) *variable component*, a mass flow rate value belonging to a cycle; and (c) *search component*, all pressure variables in the network. These components are depicted in Figure 2. The search space employed by TS is defined by the flow variables x_{ij} only because once the rates are fixed, the pressure variables are optimally found by NDP. Furthermore, we do not need to handle the entire set of flow variables, but only one per cycle. This is so because once you fix a flow rate in a cycle, the rest of the flows can be uniquely determined. Thus, a given state is represented by a vector $x = (x_{\alpha_1}, \ldots, x_{\alpha_m})$, where α_w is an arc that belongs to a selected cycle w.

Now, components of the proposed NDPTS procedure are briefly discussed.

Initial solution generation: To generate an initial solution, we use a two-phase procedure. First, a set of feasible flows are found and then an optimal set of pressures (for the fixed set of flows) is found by the NDP algorithm [1].

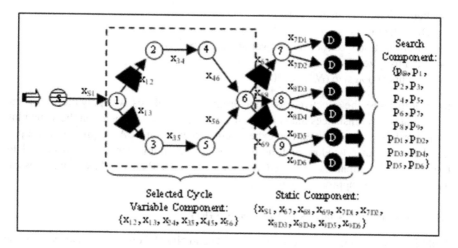

Fig. 2. Basic components of a feasible solution on a cyclic topology

Neighborhood $V(x)$: Let us define the neighborhood $V(x)$ of a given solution x. By definition, $V(x)$ is a set of solutions reachable from x via a slight modification of Δ_x units.

$$v(x) = \{x' \in R \mid x' = x \pm j\Delta_x, j = 1, 2, \ldots, Nsize/2\} \qquad (8)$$

where *Nsize* its the predefined neighborhood size. Note that, for a given solution, we do not store the entire solution but only the flow in the selected arc to be modified.

Tabu list: The Tabu List (TL) is used to keep attributes that created the best solution in past iterations so that they can not be used to create new solution candidates. As iterations proceed, a new attribute value enters in the TL and the oldest one, if it exceeds the TL size, is released. Particularly, the size of TL is the control parameter of TS. The size of TL that provided good solutions usually grows with the size of $V(x)$.

Termination Criteria: The search will terminate after *iter_max* iterations, which is a user-specified parameter.

5   Empirical Evaluation

The proposed TS was developed in C++ and run on a Sun Ultra 10 workstation under Solaris 7. All of the compressor-related data, described in Villalobos-Morales et al. [14], was provided by a consulting firm in the pipeline industry. For the tabu list size and the neighborhood size, several preliminar experiments were done using values of $\{5, 8, 10\}$ and $\{20, 30, 40\}$, respectively. Because of space constraints a full description of the fine-tuning experiment and the instances tested are available from the authors. In preliminar computations for fine-tuning the procedure we have found the following algorithmic parameters gave the best results:

- Iteration limit (*iter_max* = 100).
- Discretization size in $V(x)$ ($\Delta_x = 5$)
- Discretization size for pressure variables ($\Delta_p = 20$)
- Tabu list size (*Ttenure* = 8),
- Neighborhood size (*Nsize* = 20)

In order to assess the effectiveness of the proposed procedures, we apply the algorithms to solving several instances under different cyclic network topologies on the same platform. For this, we carried out two experiments. In experiment A we present a comparison between our procedure and the best GRG-based implementation known to date. Experiment B compares our procedure with Carter's NDP approach, which represents the best DP-based approach known to date.

5.1 Comparative Analysis 1: NDPTS vs. GRG

Table 1 shows a comparison between the GRG and NDPTS on cyclic networks. For the GRG we used the implementation in [4]. The first column shows the instances tested. Here the *ncm* suffix means that the instance has n nodes and m compressor stations. The second and third column show the GRG and NDPTS solution, respectively. The last column shows the relative improvement of NDPTS over GRG.

Table 1. Comparison between GRG and NDPTS

Instance	GRG	NDPTS	RI (%)
net-c-6c2-C1	2,312,548.24	2,288,252.53	1.05
net-c-6c2-C4	1,393,061.12	1,393,001.99	0.04
net-c-6c2-C7	1,988,998.79	1,140,097.39	42.67
net-c-10c3-C2	Not found	4,969,352.82	N/A
net-c-10c3-C4	5,610,932.12	2,237,507.93	60.12
net-c-15c5-C2	6,313,810.78	4,991,453.59	20.94
net-c-15c5-C4	3,555,353.60	3,371,985.41	5.15
net-c-15c5-C5	Not found	7,962,687.43	N/A
net-c-17c6-C1	Not found	8,659,890.72	N/A
net-c-19c7-C4	Not found	8,693,003.78	N/A
net-c-19c7-C8	Not found	7,030,280.45	N/A

First, the NDPTS was able to deliver solutions to all instances tested, whereas GRG failed for five of these. The results indicate that NDPTS procedure outperforms GRG in terms of solution quality. In terms of computational effort, GRG run in less than 2 sec. while NDPTS run in a range of 270-400 seconds.

5.2 Comparative Analysis 2: NDPTS vs. NDP

We now present a comparative analysis showing the improvement achieved by the NDPTS approach when compared with the simple NDP approach, Carter's algorithm which represents the current state-of-the-art. In Table 2, the first column shows the

Table 2. Comparison between NDP and NDPTS

Instance	NDP	NDPTS	RI (%)
net-c-6c2-C1	2,317,794.61	2,288,252.53	1.27
net-c-6c2-C4	1,394,001.99	1,393,001.99	0.07
net-c-6c2-C7	1,198,415.69	1,140,097.39	4.86
net-c-10c3-C2	6,000,240.25	4,969,352.82	17.18
net-c-10c3-C4	2,533,470.72	2,237,507.93	11.68
net-c-15c5-C2	6,006,930.42	4,991,453.59	16.90
net-c-15c5-C4	3,669,976.44	3,371,985.41	8.11
net-c-15c5-C5	8,060,452.17	7,962,687.43	1.21
net-c-17c6-C1	9,774,345.45	8,659,890.72	11.40
net-c-19c7-C4	12,019,962.22	8,693,003.78	27.67
net-c-19c7-C8	8,693,003.78	7,030,280.45	19.12

instances tested, the second column shows the solution delivered by NDP, the third column shows the best value found NDPTS, and the last column presents the relative improvement percentage of NDPTS over NDP, that is:

$$\frac{g_{NDP} - g_{NDPTS}}{g_{NDPTS}} \times 100\%$$

As can be seen, the improvement of NDPTS over the DP, is larger than 10% on 6 of 11 tested instances, and larger than 2% in 8 of the 11 instances. In only one of them the improvement is lower than 1%. The NDP runs in less than 20 sec.

5.3   A Lower Bound Comparison

To assess the quality of the solutions delivered by the algorithm it is necessary to derive a lower bound. Now, deriving lower bounds for a non-convex problem can become a very difficult task. Obtaining convex envelopes can be as difficult as solving the original problem. However, for this problem we note two important facts that lead us to an approximate lower bound. First, by relaxing constraint (4) in model FCMP the problems becomes separable in each compressor station. That is, the relaxed problem consists of optimizing each compressor station individually. Now, this is still a non-convex problem, however, we exploit the fact that in each compressor, the objective is a function of three variables only, so we build a three-dimensional grid on these three variables and perform an exhaustive evaluation for finding the global optimum of the relaxed problem (for a specified discretization).

Table 3 shows these results. The first column displays the instances tested, the second and third columns show the lower bound and the best value found by the heuristic, respectively, and the last column shows the relative optimality gap obtained by NDPTS.

As can be seen from the table, all of the tested instances have a relative optimality gap of less than 17%, 7 out of 11 instances tested have a relative gap of less than 10%, and 3 of these observed an optimality gap of less than 1%. This shows the effectiveness of the proposed approach. Finally, although our NDPTS algorithm finds better solutions than the GRG method or the simple NDP, it is more computationally expensive. In

Table 3. Solution quality

Instance	LB	NDPTS	Gap (%)
net-c-6c2-C1	2,287,470.58	2,288,252.53	0.03
net-c-6c2-C4	1,392,354.29	1,393,001.99	0.05
net-c-6c2-C7	949,909.48	1,140,097.39	16.68
net-c-10c3-C2	4,303,483.50	4,969,352.82	13.40
net-c-10c3-C4	2,015,665.98	2,237,507.93	9.91
net-c-15c5-C2	4,955,752.90	4,991,453.59	0.72
net-c-15c5-C4	3,103,697.48	3,371,985.41	7.96
net-c-15c5-C5	6,792,248.08	7,962,687.43	14.69
net-c-17c6-C1	8,129,730.11	8,659,890.72	6.12
net-c-19c7-C4	7,991,897.18	8,693,003.78	8.06
net-c-19c7-C8	5,897,768.92	7,030,280.45	16.10

general, any additional time leading to even small improvements can be easily justified since the costs involved in natural gas transportation are relatively huge.

6   Conclusions

In this work we have proposed a hybrid heuristic based on NDP and TS for a very difficult problem arising in the natural gas pipeline industry. The NDPTS implementation, based primarily in a short-term memory strategy, proved very successful in the experimental work as it was able to deliver solutions of much better quality than those delivered by earlier approaches. This represents, to the best of our knowledge, a significant contribution to the state of the art in this area of work.

There are still many areas for forthcoming research. The proposed procedure is a basic short-term memory tabu search. It would be interesting to incorporate advanced TS strategies such as intensification and diversification. In addition, one of the great challenges in the industry is to address time-dependent systems from the optimization perspective.

Acknowledgments. This research was supported by the Mexican National Council for Science and Technology (CONACYT grant J33187-A) and Universidad Autónoma de Nuevo León under its Scientific and Technological Research Support Program (UANL-PAICYT grant CA820-04). We also would like to thank two anonymous reviewers whose input helped improve the presentation of this paper.

References

1. C. Borraz-Sánchez and R. Z. Ríos-Mercado. A non-sequential dynamic programming approach for natural gas network optimization. *WSEAS Transactions on Systems*, 3(4):1384–1389, 2004.
2. R. G. Carter. Pipeline optimization: Dynamic programming after 30 years. In *Proceedings of the 30th PSIG Annual Meeting*, Denver, October 1998.

3. D. Cobos-Zaleta and R. Z. Ríos-Mercado. A MINLP model for minimizing fuel consumption on natural gas pipeline networks. In *Proceedings of the XI Latin-Ibero-American Conference on Operations Research*, Concepción, Chile, October 2002.

4. H. J. Flores-Villarreal and R. Z. Ríos-Mercado. Computational experience with a GRG method for minimizing fuel consumption on cyclic natural gas networks. In N. E. Mastorakis, I. A. Stathopulos, C. Manikopoulos, G. E. Antoniou, V. M. Mladenov, and I. F. Gonos, editors, *Computational Methods in Circuits and Systems Applications*, pages 90–94. WSEAS Press, Athens, Greece, 2003.

5. F. Glover and M. Laguna. *Tabu Search*. Kluwer, Boston, 1997.

6. R. Horst, P. M. Pardalos, and N. V. Thoai. *Introduction to Global Optimization*. Kluwer Academic Publishers, Dordrecht, The Netherlands, 1995.

7. H. S. Lall and P. B. Percell. A dynamic programming based gas pipeline optimizer. In A. Bensoussan and J. L. Lions, editors, *Analysis and Optimization of Systems*, volume 144 of *Lecture Notes in Control and Information Sciences*, pages 123–132, Berlin, 1990. Springer-Verlag.

8. A. J. Osiadacz. Dynamic optimization of high pressure gas networks using hierarchical systems theory. In *Proceedings of the 26th PSIG Annual Meeting*, San Diego, October 1994.

9. A. J. Osiadacz and M. Górecki. Optimization of pipe sizes for distribution gas network design. In *Proceedings of the 27th PSIG Annual Meeting*, Albuquerque, October 1995.

10. A. J. Osiadacz and S. Swierczewski. Optimal control of gas transportation systems. In *Proceedings of the 3rd IEEE Conference on Control Applications*, pages 795–796, August 1994.

11. P. B. Percell and M. J. Ryan. Steady-state optimization of gas pipeline network operation. In *Proceedings of the 19th PSIG Annual Meeting*, Tulsa, October 1987.

12. R. Z. Ríos-Mercado. Natural gas pipeline optimization. In P. M. Pardalos and M. G. C. Resende, editors, *Handbook of Applied Optimization*, chapter 18.8.3, pages 813–825. Oxford University Press, New York, 2002.

13. R. Z. Ríos-Mercado, S. Wu, L. R. Scott, and E. A. Boyd. A reduction technique for natural gas transmission network optimization problems. *Annals of Operations Research*, 117(1–4):217–234, 2002.

14. Y. Villalobos-Morales, D. Cobos-Zaleta, H. J. Flores-Villarreal, C. Borraz-Sánchez, and R. Z. Ríos-Mercado. On NLP and MINLP formulations and preprocessing for fuel cost minimization of natural gas transmission networks. In *Proceedings of the 2003 NSF Design, Service and Manufacturing Grantees and Research Conference*, Birmingham, January 2003.

15. Y. Villalobos-Morales and R. Z. Ríos-Mercado. Preprocesamiento efectivo de un problema de minimización de combustible en sistemas de transporte de gas natural. *Revista Ingeniería de Sistemas*, 2005. Forthcoming.

16. P. J. Wong and R. E. Larson. Optimization of natural-gas pipeline systems via dynamic programming. *IEEE Transactions on Automatic Control*, AC–13(5):475–481, 1968.

17. P. J. Wong and R. E. Larson. Optimization of tree-structured natural-gas transmission networks. *Journal of Mathematical Analysis and Applications*, 24(3):613–626, 1968.

18. S. Wu, E. A. Boyd, and L. R. Scott. Minimizing fuel consumption at gas compressor stations. In J. J.-W. Chen and A. Mital, editors, *Advances in Industrial Engineering Applications and Practice I*, pages 972–977, Cincinnati, Ohio, 1996. International Journal of Industrial Engineering.

19. S. Wu, R. Z. Ríos-Mercado, E. A. Boyd, and L. R. Scott. Model relaxations for the fuel cost minimization of steady-state gas pipeline networks. *Mathematical and Computer Modelling*, 31(2–3):197–220, 2000.

Hybrid Tabu Search for Lot Sizing Problems

João Pedro Pedroso[1] and Mikio Kubo[2]

[1] DCC-FC and LIACC, Universidade do Porto,
Rua do Campo Alegre, 823, 4150-180 Porto, Portugal
jpp@ncc.up.pt
[2] Supply Chain Optimization Lab.,
Tokyo University of Marine Science and Technology,
2-1-6 Etsuujima Koutou-ku, Tokyo 135-8533, Japan
kubo@e.kaiyodai.ac.jp

Abstract. This paper presents a hybrid tabu search strategy for lot sizing problems. This strategy allows the exploitation of the quality of the well-known relax-and-fix heuristic, inside a tabu search framework which enforces diversity.

The computational results show an advantage of this strategy when compared to a version of the relax-and-fix heuristic and to time constrained branch-and-bound.

1   Introduction

Lot sizing is a class of combinatorial optimization problems with applications in production planning. In these problems there is a fixed number of periods, and in each period production of items can occur in machines. A machine has to be appropriately setup for being able to produce, and this setup implies, e.g., the payment of a fixed cost, or the reduction of the machine working time by some fixed amount.

The amount produced in a given period can be used to satisfy the demand of that period, or remain in inventory. When production can also be used to satisfy demand of preceding periods, the models are said to allow backlogging.

Lot sizing problems can be classified into small-bucket or big-bucket models. On small bucket models, each machine can produce at most one item per period; on big-bucket models, several items can be manufactured in each period.

Good surveys on lot sizing are provided in [3] and [6].

2   The Lot Sizing Model

The problem that we deal with in this paper is a lot sizing problem belonging to the big bucket class: more than one setup is allowed per period, as long as the machine capacities are respected.

The costs that are to be taken into account are setup costs, variable production costs, and inventory and backlog costs. Unitary values for each of them can vary from period to period.

M.J. Blesa et al. (Eds.): HM 2005, LNCS 3636, pp. 66–77, 2005.

The decision variables in this lot sizing problem concern the manufacture or not of a product in each period, as well as the amount to produce. The setup, binary variable y_{pmt} is 1 if product p is manufactured in machine m during period t, and 0 otherwise. The continuous variable x_{pmt} is the corresponding manufactured amount.

Let T be the number of periods and $\mathcal{T} = \{1, \ldots, T\}$. Let $\mathcal{P}$ be the set of products and $\mathcal{M}$ be the set of machines. Let furthermore $\mathcal{M}^p$ be the subset of machines that are compatible with the production of p. The setup costs are then determined by:

$$F = \sum_{p \in \mathcal{P}} \sum_{m \in \mathcal{M}^p} \sum_{t \in \mathcal{T}} f_{pmt}\, y_{pmt}, \tag{1}$$

where f_{pmt} is the cost of setting up machine m on period t for producing p. Similarly, variable costs are

$$V = \sum_{p \in \mathcal{P}} \sum_{m \in \mathcal{M}^p} \sum_{t \in \mathcal{T}} v_{pmt}\, x_{pmt}, \tag{2}$$

where v_{pmt} is the variable cost of production of p on machine m, period t. If h_{pt} is the amount of product p that is kept in inventory at the end of period t, the inventory costs can be determined by

$$I = \sum_{p \in \mathcal{P}} \sum_{t \in \mathcal{T}} i_{pt}\, h_{pt}, \tag{3}$$

where i_{pt} is the unit inventory cost for product p on period t. Analogously, if g_{pt} is the amount of product p that failed to meet demand at the end of period t, the backlog costs can be determined by

$$B = \sum_{p \in \mathcal{P}} \sum_{t \in \mathcal{T}} b_{pt}\, g_{pt}, \tag{4}$$

where b_{pt} is the unit backlog cost for product p on period t. The lot sizing objective can now be written as

$$\text{minimise } z = F + V + I + B. \tag{5}$$

If the demand of a product p in period t is D_{pt}, the flow conservation constraints can be written as

$$h_{p,t-1} - g_{p,t-1} + \sum_{m \in \mathcal{M}^p} x_{pmt} = D_{pt} + h_{pt} - g_{pt} \quad \forall\, p \in \mathcal{P}, \forall\, t \in \mathcal{T}. \tag{6}$$

The initial inventory and backlog for each product p should be assigned to h_{p0} and g_{p0}, respectively (and possibly equivalent assignments might be made for h_{pT} and g_{pT}).

There is a limit on the time that each machine is available on each period; this implies that

$$\sum_{p \in \mathcal{P}: m \in \mathcal{M}^p} \left(\frac{x_{pmt}}{\gamma_{pm}} + \tau_{pmt}\, y_{pmt} \right) \leq A_{mt} \quad \forall\, m \in \mathcal{M}, \forall\, t \in \mathcal{T}. \tag{7}$$

In this equation, γ_{pm} is the total capacity of production of product p on machine m per time unit, τ_{pmt} is the setup time required if there is production of p on machine m during period t, and A_{mt} is the number of time units available for production on machine m during period t.

Manufacturing of a given product can only occur on machines which have been setup for that product:

$$x_{pmt} \leq \gamma_{pm} A_{mt} y_{pmt} \quad \forall p \in \mathcal{P}, \forall m \in \mathcal{M}^p, \forall t \in \mathcal{T}. \tag{8}$$

The problem can be summarized as the following mixed-integer program (MIP):

minimise $z = F + V + I + B$

subject to : $F = \sum_{p \in \mathcal{P}} \sum_{m \in \mathcal{M}} \sum_{t \in \mathcal{T}} f_{pmt} y_{pmt}$

$V = \sum_{p \in \mathcal{P}} \sum_{m \in \mathcal{M}} \sum_{t \in \mathcal{T}} v_{pmt} x_{pmt}$

$I = \sum_{p \in \mathcal{P}} \sum_{t \in \mathcal{T}} i_{pt} h_{pt}$

$B = \sum_{p \in \mathcal{P}} \sum_{t \in \mathcal{T}} b_{pt} g_{pt}$

$h_{p,t-1} - g_{p,t-1} + \sum_{m \in \mathcal{M}^p} x_{pmt} = D_{pt} + h_{pt} - g_{pt}, \quad \forall p \in \mathcal{P}, \forall t \in \mathcal{T}$

$\sum_{p \in \mathcal{P}: m \in \mathcal{M}^p} \left(\dfrac{x_{pmt}}{\gamma_{pm}} + \tau_{pmt} y_{pmt} \right) \leq A_{mt}, \quad \forall m \in \mathcal{M}, \forall t \in \mathcal{T}$

$x_{pmt} \leq \gamma_{pm} A_{mt} y_{pmt} \quad \forall p \in \mathcal{P}, \forall m \in \mathcal{M}^p, \forall t \in \mathcal{T}$

$F, V, I, B \in \mathbb{R}^+$
$h_{pt}, g_{pt} \in \mathbb{R}^+, \quad \forall p \in \mathcal{P}, \forall t \in \mathcal{T}$
$x_{pmt} \in \mathbb{R}^+, y_{pmt} \in \{0, 1\}, \quad \forall p \in \mathcal{P}, \forall m \in \mathcal{M}, \forall t \in \mathcal{T}$

$\hspace{11cm} (9)$

3   Construction: Relax-and-Fix-One-Product

For the construction of a solution to the problem defined by Problem 9, we consider partial relaxations of the initial problem, in a variant of the classic relax-and-fix [8,9] heuristic.

In the basic form of the relax-and-fix heuristic, each period is treated independently. The strategy starts by relaxing all the variables except those of period 1, thus keeping y_{pm1} integer and relaxing integrity for all other y_{pmt}. This MIP is solved, determining the heuristic values for variables $\bar{y}_{pm1}$ (i.e., the binary variables of the first period). The approach then moves to the second period. The variables of the first period are fixed at $y_{pm1} = \bar{y}_{pm1}$, the variables y_{pm2} are integer, and all the other y_{pmt} relaxed; this determines the heuristic value for y_{pm2}. These steps are repeated, until all the y variables are fixed, as described in Algorithm 1.

Algorithm 1: Relax-and-fix heuristic.

RELAXANDFIX()
(1)   relax $y_{pmt}, \forall\, p \in \mathcal{P},\ \forall\, m \in \mathcal{M},\ \forall\, t \in \mathcal{T}$, as continuous variables
(2)   **for** $t = 1$ **to** T
(3)   **foreach** $p \in \mathcal{P}$
(4)   **foreach** $m \in \mathcal{M}^p$
(5)   set y_{pmt} as integer
(6)   solve Problem 9, determining $\bar{y}_{pmt},\ \forall p \in \mathcal{P},\ \forall m \in \mathcal{M}^p$
(7)   **foreach** $p \in \mathcal{P}$
(8)   **foreach** $m \in \mathcal{M}^p$
(9)   fix $y_{pmt} := \bar{y}_{pmt}$
(10)   **return** $\bar{y}$

This approach is reported to provide very good solutions for many lot sizing problems. However, for large instances the exact MIP solution of even a single period can be too time consuming. Therefore, we propose a variant were each MIP determines only the variables of one period *that concern a single product*. We call this approach *relax-and-fix-one-product*; it is outlined in Algorithm 2 (were a random permutation of a set $\mathcal{P}$ is denoted by $\mathscr{R}(\mathcal{P})$).

Algorithm 2: Relax-and-fix-one-product variant.

RELAXANDFIXONEPRODUCT()
(1)   relax $y_{pmt}, \forall\, p \in \mathcal{P},\ \forall\, m \in \mathcal{M},\ \forall\, t \in \mathcal{T}$, as continuous variables
(2)   **for** $t = 1$ **to** T
(3)   **foreach** $p \in \mathscr{R}(\mathcal{P})$
(4)   **foreach** $m \in \mathcal{M}^p$
(5)   set y_{pmt} as integer
(6)   solve Problem 9, determining $\bar{y}_{pmt},\ \forall m \in \mathcal{M}^p$
(7)   **foreach** $m \in \mathcal{M}^p$
(8)   fix $y_{pmt} := \bar{y}_{pmt}$
(9)   **return** $\bar{y}$

3.1 Solution Reconstruction

As we will see in the next section, the relax-and-fix-one-product construction mechanism can be interestingly used in a different context: that of completing a solution that has been partially destructed. For this purpose, all that is required is to check if incoming $\bar{y}_{pmt}$ variables are initialized or not; if they are initialized, they should be fixed in the MIP at their current value. Otherwise, they are treated as in previous algorithm: their are made integer if they belong to the period and product currently being dealt, and relaxed otherwise. These ideas are described in Algorithm 3 (we denote a random permutation of a set $\mathcal{P}$ by $\mathscr{R}(\mathcal{P})$).

Algorithm 3: Relax-and-fix for solution reconstruction

RECONSTRUCT($\bar{y}$)

```
(1)      for t = 1 to T
(2)          foreach p ∈ P
(3)              foreach m ∈ M^p
(4)                  if ȳ_pmt is not initialized
(5)                      relax y_pmt
(6)                  else
(7)                      fix y_pmt := ȳ_pmt
(8)      for t = 1 to T
(9)          foreach p ∈ ℛ(P)
(10)             U := {}
(11)             foreach m ∈ M^p
(12)                 if ȳ_pmt is not initialized
(13)                     set y_pmt as integer
(14)                     U := U ∪ {(p, m, t)}
(15)                 solve Problem 9, determining ȳ_pmt, ∀(p, m, t) ∈ U
(16)                 foreach (p, m, t) ∈ U
(17)                     fix y_pmt := ȳ_pmt
(18)     return ȳ
```

4   A Hybrid Tabu Search Approach

In this section we present a hybrid metaheuristic approach devised for tackling the lot sizing problem. The approach is a two-fold hybrid, were relax-and-fix-one-product is used to initialize a solution, or complete partial solutions, and tabu search [2] is responsible for creating diverse points for restarting relax-and-fix. Before each restart, the current tabu search solution is partially destructed; its reconstruction is made by means of the relax-and-fix-one-product procedure presented in Algorithm 3.

4.1   Solution Representation and Evaluation

In what concerns tabu search, the subset of variables of the Problem 9 which is necessary to store is the set of y_{pmt} variables; all the continuous variables can be determined in function of these. Thus, a tabu search solution will consist of a matrix of the $\bar{y}_{pmt}$ binary variables.

The evaluation of a solution can be made through the solution of the Problem 9, with all the binary variables fixed at the values $\bar{y}_{pmt}$. As all the binary variables are fixed, this problem is a linear program (LP). The value of z at the optimal solution of this LP will provide the evaluation of the quality of $\bar{y}_{pmt}$. The values of all the other variables x, h and g corresponding to $\bar{y}_{pmt}$ are also determined through this LP solution.

4.2   Tabu Search

The tabu search framework, presented in Algorithm 4, is based only on short term memory. This procedure has a parameter, *tlim*, which is the limit of CPU to be used in the search (as an alternative to the usual stopping criterion, based on the number of iterations). The remaining arguments are a seed for initializing the random number generator, and the name of the instance to be solved.

Algorithm 4: Hybrid tabu search.

TABUSEARCH(*tlim, seed, instance*)

```
(1)     store instance information as global data T, P, M, f, g, ...
(2)     initialize random number generator with seed
(3)     ȳ := RELAXANDFIXONEPRODUCT()
(4)     ȳ* := ȳ
(5)     n := |T| × |P|
(6)     Θ := ((−n, ..., −n), ..., (−n, ..., −n))
(7)     i := 1
(8)     while CPUTIME() < tlim
(9)         ȳ := TABUMOVE(ȳ, ȳ*, i, Θ)
(10)        if ȳ is better than ȳ*
(11)            ȳ* := ȳ
(12)        i := i + 1
(13)    return ȳ*
```

4.3 Neighborhood and Candidate Selection

In the course of a tabu search iteration, the neighborhood of the current solution is searched as presented in sAlgorithm 5. The arguments of this algorithm are the current solution $\bar{y}$, the best solution found $\bar{y}^*$, the current iteration i, and the tabu matrix Θ.

Tabu information is kept in the matrix Θ, where Θ_{pm} holds the iteration at which some variable y_{pmt} has been updated. The tabu tenure is a random value, drawn in each iteration between 1 and the number of integer variables and stored in the variable d, on line 6 of Algorithm 5 ($\mathscr{R}[a, b]$ is the notation used for a random integer with uniform distribution in $[a, \ldots, b]$). If the current iteration is i, a move involving product p and machine m will be tabu if $i - \Theta_{pm} \leq d$; otherwise (i.e., if $i - \Theta_{pm} > d$) it is not tabu. Making the tabu tenure a random value simplifies the parameterization of the algorithm.

The neighborhood used consists of solutions where manufacturing a product in a given period and machine is stopped, and its manufacture is attempted in different machines, on the same period. Hence, for a given solution y we start by checking, in a random order, what are the products which are being manufactured in the first period. Let us suppose that a product p is being manufactured in machine m, i.e., $y_{pm1} = 1$. The first neighbor is a solution where $y_{pm1} = 0$, all the other elements being equal to their equivalent in y. Other neighbors have $y_{pm'1} = 1$ for all the machines $m' \neq m$ where p was *not* being produced. After checking the first period, we check sequentially the periods $2, \ldots, T$, as detailed in lines 2 to 20 of Algorithm 5. This is, therefore, a composed neighborhood, where one or two moves are allowed. On line 3, $\mathscr{R}(\mathcal{P})$ is the notation used for a random permutation of the set of products $\mathcal{P}$.

Algorithm 5: Move during each tabu search iteration

TABUMOVE($\bar{y}, \bar{y}^*, i, \Theta$)

(1) $\bar{y}' := \bar{y}$
(2)   **for** $t = 1$ **to** T
(3)   **foreach** $p \in \mathscr{R}(\mathcal{P})$
(4) $\mathcal{S} := \{m \in \mathcal{M}^p : \bar{y}_{pmt} = 1\}$
(5) $\mathcal{U} := \{m \in \mathcal{M}^p : \bar{y}_{pmt} = 0\}$
(6) $d := \mathscr{R}[1, |\mathcal{P}| \times |\mathcal{M}| \times |\mathcal{T}|]$
(7)   **foreach** $m \in \mathcal{S}$
(8)   fix $\bar{y}_{pmt} := 0$
(9)   **if** $\bar{y}$ is better than $\bar{y}^*$ **or** $(i - \Theta_{pm} > d$ **and** $\bar{y}$ is better than $\bar{y}')$
(10)   **return** $\bar{y}$
(11)   **if** $i - \Theta_{pm} > d$ **and** $(\bar{y}^c$ is not initialized **or** $\bar{y}$ is better than $\bar{y}^c)$
(12) $\bar{y}^c := \bar{y}, m_1 := (p, m, t)$
(13)   **foreach** $m' \in \mathcal{U}$
(14)   fix $\bar{y}_{pm't} := 1$
(15)   **if** $\bar{y}$ is better than $\bar{y}^*$ **or** $(i - \Theta_{pm} > d$ **and** $\bar{y}$ is better than $\bar{y}')$
(16)   **return** $\bar{y}$
(17)   **if** $i - \Theta_{pm} > d$ **and** $(\bar{y}^c$ is not initialized **or** $\bar{y}$ is better than $\bar{y}^c)$
(18) $\bar{y}^c := \bar{y}, m_1 := (p, m, t), m_2 := (p, m', t)$
(19)   restore $\bar{y}_{pm't} := 0$
(20)   restore $\bar{y}_{pmt} := 1$
(21) $\alpha := \mathscr{R}$
(22)   un-initialize $\alpha\%$ of the $\bar{y}^c$ variables
(23)   **if** $\bar{y}^c$ is not initialized
(24)   select a random index (p, m, t)
(25) $\bar{y}^c := \bar{y}, \bar{y}^c_{pmt} := 1 - \bar{y}_{pmt}, \Theta_{pm} := i$
(26)   **else**
(27) $(p, m, t) := m_1, \Theta_{pm} := i$
(28)   **if** m_2 is initialized
(29) $(p, m, t) := m_2, \Theta_{pm} := i$
(30) $\bar{y} :=$ RECONSTRUCT$(\bar{y}^c)$
(31)   **return** $\bar{y}$

If a neighbor improving the best solution could be found, it is returned immediately. A neighbor solution is immediately returned also if it is not tabu and it improves the input solution (lines 9–10 and 15–16).

In the case that no improving move could be found in the whole neighborhood, we force a diversification: the solution is partially destructed, as shown in lines 21 and 22 ($\mathscr{R}$ is the notation used for a random real number with uniform distribution in $[0, 1]$).

The best found move is then applied and made tabu (lines 27 to 29), and the solution is reconstructed (line 30). Notice that this move is excluded from the parts of the solution that are to be reconstructed.

Lines 23 to 25 prevent the case where the search is blocked, all moves being tabu; in such a case, a random move is taken.

5   Computational Results

The implementation of the hybrid tabu search is somewhat tricky, as it involves the interaction with LP and MIP solvers; but for the MIP class of problems there is no alternative. The programs were implemented in the Python language [7], making use of an interface to the GLPK optimization library [4]. The programming the code was written in just a few pages of code; it is available, together with the mathematical programming model, in [5].

5.1   Practical Benchmarks

The strategies described in this paper were tested on a series of benchmark instances. These instances have a random component, but are derived from a real-world problem, where there are 12 products involved and the number of periods is 12. We have produced an instance based on this full-size problem, and smaller instances by reducing the number of periods and randomly selecting a subset of products. The machines made available on smaller instances are those compatible with the selected products, and the demand is randomly drawn, the average being the true estimated demand.

The characteristics of the selected instances are presented in table 1; these instances are available for downloading in [5].

Table 1. Names and characteristics of the instances used for benchmarking

Name	Number of periods	Number of products	Number of integers	Number of variables	Number of constraints
fri-02	2	2	20	56	45
fri-05	5	5	135	334	235
fri-07	7	7	210	536	369
fri-09	9	9	306	796	554
fri-12	12	12	492	1300	857

The results obtained both by the branch-and-bound solver available in the GLPK kit and by the metaheuristics presented in this paper are presented in table 2. In both the situations, the search was limited to a CPU time of one hour in a computer running Linux 2.4.27-2-686, with an Intel Pentium 4 processor at 1.6 GHz.

Notice that as there is a random component on the relax-and-fix-one-product heuristic, the solutions found can be different from run to run. However, due to the structure of the costs on these instances, the solution is virtually always the same.

5.2   Algorithm Behavior

For providing a deeper insight on the behavior of the algorithm, we also present graphics of the evolution of the current solution with respect to the number of

Table 2. Results obtained by the relax-and-fix-one-product heuristic, by the hybrid tabu search, and by time-constrained branch-and-bound. Branch-and-bound and the hybrid tabu search were limited to 3600 seconds of CPU time for each instance. (* *fri-02* was solved by branch-and-bound to optimality in less that one second.)

| Name | Relax-and-fix (average) | | Hybrid tabu search sol. | | | branch-and- |
	time (s)	solution	worst	average	best	bound best sol.
fri-02	< 1	13.897	13.897	13.897	13.897	13.897*
fri-05	1.4	49.904	48.878	48.878	48.878	48.878
fri-07	2.3	131.095	126.865	126.197	126.030	127.604
fri-09	4.6	213.405	209.201	208.303	207.640	235.125
fri-12	12.4	277.451	275.004	274.681	273.963	431.660

iterations, for a typical run. We have selected the instance *fri-12*, and analyzed the evolution of the objective value of the current solution, z, as well as that of the best solution, z^*. To illustrate the importance of the destruction/reconstruction phase, we have plotted in figure 1 this evolution for the hybrid tabu search algorithm, and also for pure tabu search, without that phase.

These graphics show the importance of the destruction/reconstruction phase, as without it tabu search quickly moves from the initial (relax-and-fix-one-product) solution into poor areas, and cannot easily reach good solutions again. This is due to the fact that a large number of moves is required to change a good solution into another good solution. Partial destroying the solution and reconstructing it with the relax-and-fix-one-product heuristic can do the large number of changes in the solution that is required to bring it to good places. This phase is used whenever tabu search cannot find an neighbor improving the current solution; if improving neighbors are found, the destruction/reconstruction cycle is skipped.

On this run of the hybrid metaheuristic the number of iterations allowed in the 3600 seconds CPU time was 178; on 164 of these the solution was destructed and reconstructed with relax-and-fix-one-product. There were 10 improvements on the best solution found; 1 was found stopping production in a machine on the current solution, 7 were found starting production in a different machine, and 2 were found on reconstruction. Although these values vary widely from run to run and with the instance, in general there can be observed (as in this case) solution improvements on both the neighborhood search and the destruction/reconstruction phase.

5.3   Other Benchmarks

Although the algorithm was designed for a specific application, as it encompasses most of the relevant issues in big bucket lot sizing, it can be used with no modification for other applications. We have selected some of the relevant (big-bucket) benchmarks from the LOTSIZELIB [10], adapted them to allow backlog (though at a very high penalty) and used hybrid tabu search to solve these instances on the same computational setup used in section 5.1.

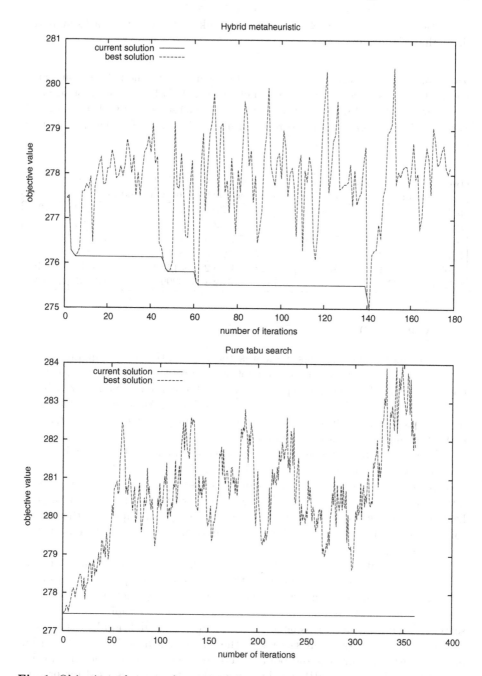

Fig. 1. Objective value as a function of the iteration number for the *fri-12*instance. The top graphic corresponds to the hybrid meta-heuristic; the bottom one corresponds to pure tabu search, without the destruction/reconstruction steps. Both algorithms run 3600 seconds.

Notice that by allowing backlog, we are making these benchmarks more difficult than the original ones.

Table 3. Results obtained by the relax-and-fix heuristic, by the hybrid tabu search, and by time-constrained branch-and-bound on LOTSIZELIB instances. Branch-and-bound and the hybrid tabu search were limited to 3600 seconds of CPU time for each instance. Optimal solutions are as reported in [1]. (* indicate best known solutions, optimality is not proven on these cases.)

Name	Relax-and-fix (average)		Hybrid tabu search sol.			branch-and-	optimal
	time (s)	solution	worst	average	best	bound best sol.	solution
pp08a	<1	7638.0	7380	7374	7360	7350	7350
rgna	<1	82.2	82.2	82.2	82.2	82.2	82.2
set1ch	13.4	56024.3	55243.5	55089.6	54950	60517.7	54537
tr6-15	1.3	40767.6	38357	38238	38054	39388	37721
tr6-30	5.1	67057.0	63422	63246.2	63132	63711	61746*
tr12-30	69.1	143014.0	137371	136762.8	136299	1940337	130599*

6   Conclusion

The main motivation for this work was the exploitation of the quality of the well-known relax-and-fix heuristic for lot sizing problems, in a setup which enforced diversity. This setup was provided by a tabu search mechanism, which was responsible for imposing some changes on the solution. After these changes were made, a part of the solution (not involving the latest changes) was destructed, and relax-and-fix was used to rebuild it.

The reason why this was required as a complement to tabu search is that non-improving moves made by tabu search rapidly force the solution into rather poor regions, because a large number of moves is required to change a good solution into another good solution. These "moves" were done by the relax-and-fix heuristic whenever tabu search could not find an neighbor improving the current solution. When improving neighbors were found, the destruction/reconstruction cycle were skipped.

The computational results obtained with hybrid tabu search on a series of benchmarks show a clear advantage of this strategy, as compared to the simple relax-and-fix-one-product heuristic and to time-limited branch-and-bound.

This work has raised several issues, which remain as topics for future research. The first is the assessment of the quality of the hybrid algorithm using a specialized branch-and-cut system, as the one provided in [1], for the solution of MIPs, instead of branch-and-bound. Another open question concerns limiting the CPU used on each MIP solution; as most of the CPU is used for proving optimality (which is not required in this context), limiting it would probably lead to significant improvements.

The tabu search framework was designed with only short term memory. This provided solutions which are good enough for the practical application to which the algorithm was designed, but if a deeper search is required it might be useful to implement more sophisticated methods, including long term memory and periodic restart from elite solutions.

References

1. G. Belvaux and Laurence A. Wolsey. Modelling issues and a specialized branch-and-cut system bc-prod. Discussion Paper 9849, Center for Operations Research and Econometrics, Université Catholique de Louvain, Louvain-la-Neuve, Belgium, 1998.
2. F. Glover and M. Laguna. *Tabu Search*. Kluwer Academic Publishers, Boston, 1997.
3. R. Kuik and M. Salomon. Batching decisions: Structure and models. *European Journal and Operational Research*, 75:243–260, 1994.
4. Andrew Makhorin. *GLPK – GNU Linear Programming Kit*. Free Software Foundation, http://www.gnu.org, 2005. Version 4.8.
5. João P. Pedroso. Hybrid tabu search for lot sizing problems: an implementation in the Python programming language. Internet repository, version 0.1, 2005. http://www.ncc.up.pt/~jpp/lsize.
6. Yves Pochet and Laurence A. Wolsey. Algorithms and reformulations for lot sizing problems. In *DIMACS Series in Discrete Mathematics and Theoretical Computer Science*, volume 20, pages 245–293, 1995.
7. Guido van Rossum et al. *Python Documentation*. PythonLabs, http://www.python.org, 2005. Release 2.3.5.
8. Mathieu Van Vyve and Yves Pochet. General heuristics for production planning problems. *INFORMS Journal on Computing*, 16:316–32, 2004.
9. Laurence Wolsey. *Integer Programming*. John Wiley & Sons, 1998.
10. Laurence A. Wolsey. LOTSIZELIB. Internet repository, version 3.0, 1996. http://www.core.ucl.ac.be/wolsey/lotsizel.htm.

Fast Ejection Chain Algorithms for Vehicle Routing with Time Windows*

Herman Sontrop[1], Pieter van der Horn[1], and Marc Uetz[2]

[1] Philips Research Laboratories, Prof. Holstlaan 4, 5656 AA Eindhoven,
The Netherlands
{Herman.Sontrop, Pieter.van.der.Horn}@philips.com
[2] Maastricht University, Quantitative Economics, P.O. Box 616,
6200 MD Maastricht, The Netherlands
M.Uetz@ke.unimaas.nl

Abstract. This paper introduces a new algorithm, based on the concept of ejection chains, to effectively target vehicle routing problems with time window constraints (VRPTW). Ejection chains create powerful compound moves within Local Search algorithms. Their potential to yield state of the art algorithms has been validated for the traveling salesman problem (TSP), for example. We show how ejection chains can be used to tackle the more general VRPTW as well. The yardstick behind ejection chain procedures is the underlying reference structure; it is used to coordinate the moves that are available for the Local Search algorithm via a given set of transition rules. Our main contribution is the introduction of a new reference structure, generalizing reference structures previously suggested for the TSP. The new reference structure, together with a set of simple transition rules, is tailored to handle the asymmetric aspects in a VRPTW. We use Tabu Search for the generation of the ejection chains, and on a higher algorithmic level, the ejection chain process is embedded into an Iterated Local Search algorithm. Computational results confirm that this approach leads to very fast algorithms, showing that ejection chain algorithms have the potential to compete with state of the art algorithms for the VRPTW.

1 Introduction

Recently, it has been shown that so-called *Stem & Cycle ejection chain* procedures can compete with state of the art implementations of the famous Lin-Kernighan algorithm [10] for solving large scale traveling salesman problems; see, e.g., [5]. This is remarkable, since Lin-Kernighan type algorithms had dominated this field for the last decades. Ejection chain procedures explicitly identify a so-called *reference structure*. This is a structure similar to, but slightly different from a solution, for example by violating certain types of constraints. Via a set of predefined *transition rules*, moves are generated from feasible solutions to

---

* This research was performed on behalf of the CENTRE FOR QUANTITATIVE METHODS, CQM BV, P.O. Box 414, 5600 AK, Eindhoven, The Netherlands.

reference structures, from one reference structure to another, and back from reference structures to solutions. This way the reference structure, together with the transition rules, define the moves that are available for a Local Search algorithm.

We address the vehicle routing problem with time window constraints (VRP-TW). Given is a number of customers in the plane, with demands, a given service or delivery time, and a fleet of identical vehicles with known limited capacities. We are asked to find a set of routes starting and ending at a central depot, such that each client is served by exactly one vehicle. Clearly, any route must not violate the capacity constraints of the vehicle. In addition, each client must be serviced within its so-called *time window*. The time window specifies an earliest and a latest time at which the delivery must begin. If a vehicle arrives at a customer before the opening of the time window, the vehicle will have to wait. Arriving after the end of the time window is not allowed. Two different solutions with the same number of vehicles are usually ranked by the total distance travelled by the vehicles (sometimes, also the waiting time is taken into account). The objective considered in this paper is the total distance travelled by the vehicles, utilizing as many vehicles as required. We opted for this objective in order to be able to compare our results to known optimal solutions, which have been obtained using the same objective. In addition, we also experimented with a slightly modified approach where the usage of vehicles is penalized, in order to primarily drive down the number of vehicles[1].

The VRPTW has been extensively studied. The best *exact* procedures can still only handle small instances, small being in the order of 50–100 customers [16]. Meta-heuristic procedures mostly minimize the number of vehicles, and among solutions with the same number of vehicles, prefer those with small total distance travelled. Since the overall literature is extensive, we refer to [16] for a thorough introduction into vehicle routing in general, and many references. For concepts of Tabu Search, Simulated Annealing and other meta-heuristics, we refer to Aarts and Lenstra [1]. As a matter of fact, Tabu Search based procedures are the majority among the most effective algorithms for VRPTW, see, e.g., [14,17,9], but also other meta-heuristic frameworks proved to be effective; see [16].

The starting point of our research, motivated by practical interest, was the idea to generalize state of the art algorithms for the traveling salesman problem to the more general vehicle routing problem with time windows, in order to obtain good solutions very quickly. Previous research on the TSP has shown that Local Search algorithms –both Lin-Kernighan (LK) algorithms and Stem & Cycle (SC) ejection chain algorithms– are very effective when the available moves include the reversal of (sub)paths. However, this is generally not true for settings with time window constraints, because path reversals introduce time window violations. In this respect, it was already suggested by Glover in [7] that an augmentation of the SC reference structure, the *Doubly Rooted* (DR) reference structure, is more suitable for the asymmetric TSP.

---

[1] Clearly, these objectives are correlated. When the capacities of the vehicles are large, however, solutions with less vehicles may lead to a larger total distance.

Departing from this observation, our main contribution is a new reference structure for the VRPTW, generalizing Glovers DR reference structure for the TSP, that particularly targets the asymmetric nature of the VRPTW by avoiding path reversals. This reference structure lies at the heart of an ejection chain procedure that is based on Tabu Search. In addition to providing a new reference structure, a novelty in our contribution is also a study of different meta-heuristics that are used to steer the generation of ejection chains. (So far, ejection chain studies have mainly focused on the lower level of control, the ejection chain process itself.) It turned out that Iterated Local Search performed extremely well in terms of speed and solution quality. Our computational results on established standard test sets from Solomon [15] confirm that the resulting hybrid meta-heuristic is indeed fast and effective.

2   Ejection Chains and Reference Structures

Most heuristics directly move from one solution to another. A different approach is to first move to intermediate structures, *reference structures*, before moving to another solution. In such procedures a certain amount of infeasibility is introduced to the initial solution, which has to be 'ejected' in order to end up with a new feasible solution, usually called a *trial solution* in this context. The ejection of infeasibility can be delayed by moving to other reference structures first, creating a chain effect. At each level of such a chain, trial solutions are available by ejecting the infeasibility. Hence the term *ejection chain*. In the words of Glover [7], 'ejection chain procedures are based on the notion of generating compound sequences of moves, leading from one solution to another, by linked steps in which changes in selected elements cause other elements to be "ejected from" their current state, position or value assignment'. The yardstick behind such ejection chains is the underlying reference structure, which is a structure that resembles a solution, but is infeasible with respect to some of the constraints. The key idea is to introduce infeasibility by allowing certain vertices to have odd degrees, something that is not possible in a feasible VRP solution. The general structure of an ejection chain is depicted in Figure 1. In the example we see an ejection chain of five levels. Starting from a solution S we introduce a certain amount of infeasibility via move 1 resulting in reference structure R_1. From R_1 we can either eject the sustained infeasibility (move 1*) or replace it by some other infeasibility (move 2). Obviously such a construction can be repeated, creating a chain. One can a priori set a maximum depth (in this case 5) and select

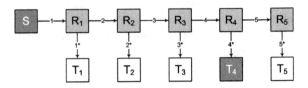

Fig. 1. Ejection chain example

the best trial solution seen in the chain. In this example the resulting compound move implicated by the chain consists of moves 1–2–3–4–4*.

2.1   The Constrained Doubly Rooted Reference Structure

It can be considered both a strength and a weakness of state of the art TSP procedures that certain types of moves require a path reversal, i.e., the reversal of a sub path of the tour. In the absence of (tight) time windows, path reversals are an important building block of powerful moves. In the presence of time windows, however, path reversals often violate time windows, especially if the time windows are tight. Therefore, our idea to create an ejection chain procedure for the VRPTW is to have a reference structure that still provides a strong connectivity between solutions, but does not rely on path reversals in order to generate moves.

The reference structure we propose supersedes both the stem & cycle (SC) reference structure and the doubly rooted (DR) reference structure of Glover [7]. Figure 2 shows an instance of the SC structure. It is a spanning subgraph that consists of one cycle and a path. The path is called the *stem*, the end of the stem is called the *tip*. The vertex that is shared by the cycle and the stem is called the *root*. Note that the root and the tip have odd degrees, hence the reference structure does not represent a feasible solution to the problem. Figure 3 shows

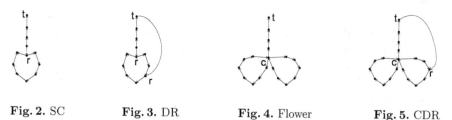

Fig. 2. SC   **Fig. 3.** DR   **Fig. 4.** Flower   **Fig. 5.** CDR

a DR reference structure. It may be conceived as arising from an SC by adding an arc (t, j) which connects the tip to an arbitrary node j on the cycle. Again, this results in a structure that has two vertices that have odd degrees. In [7] Glover refers to both these vertices as *roots*. Glover shows that the DR structure has several advantages over the SC structure. He proves a connectivity result showing that the DR allows for more direct trajectories between solutions. Most important is that the connectivity result also holds for asymmetric problems. This is the main reason to base our new reference structure on the DR concept. In addition, the DR structure provides access to moves unavailable to an SC structure. Rego [13] generalizes the SC into the *Flower* reference structure[2], depicted in Figure 4. As can be seen it is an SC structure joined with multiple cycles. The intersection of the cycles is called the *core*. An SC structure can also be seen as a (trivial) Flower. The Flower concept proved to be very fruitful for VRP problems without time windows. The reference structure we propose is closely related to both the DR structure and the Flower concept. It may be

---

[2] Rego [13] refers to the vertex t as the *root*. We use the term *tip* as in [7], however.

conceived from a Flower by adding an arc (t, j) which connects the tip to an arbitrary node j on a cycle, and is shown in Figure 5. We call it a *Constrained Doubly Rooted* reference structure (CDR). We use the word 'constrained' since the core, the depot vertex, always represents one of the roots. In Section 5 we briefly discuss an even more general, unconstrained version of the CDR. Note that a CDR, too, has two vertices that have odd degrees. We will refer to the vertex that represents the depot as the *core*, denoted by c. The other vertex with an odd degree is called the *root*, denoted by r. A vertex v such that the arc (v, r) exists is called a *subroot*. Note that when we eject such an arc (v, r) we obtain the Flower structure, therefore v could also be called an *implicit tip*.

2.2 Transition Rules

In order to exploit the reference structure, we need three types of *transition rules*. *S*-rules generate a CDR from a given start solution, *E*-rules generate one CDR from another, and *T*-rules generate a trial solution from a CDR. We state five simple transition rules that turn out to be sufficient.

S1-rule (Solution to CDR) Eject (s, c) Add (s, j) where $j \neq c$ and j not on path $c \to s$	 $(s = 4, j = 2)$
S2-rule (Solution to CDR) Add (c, p) where $p \neq c$ Eject (q, p) where q is the predecessor of p Add (q, j) where $j \neq c$ and $j \neq p$ and j not on path $c \to q$	 $(p = 1, q = 4, j = 2)$
E1-rule (CDR to CDR) Eject (s, r) where $s \neq c$ Add (s, j) where $j \neq r$ and $j \neq c$ and j not on path $c \to s$	 $(s = 4, r = 2, j = 3)$
T1-rule (CDR to Solution) Eject (s, r) where $s \neq c$ Add (s, j) where $s \neq j$ Eject (c, j)	 $(s = 4, r = 2, j = 1)$
T2-rule (CDR to Solution) Eject (s, r) Add (s, c)	 $(s = 1, r = 2)$

None of these rules involves a path reversal. The moves, however, can easily be adapted to include path reversals, if desired. In the figures, the core c is always the lowest vertex, and the root, if present, is displayed shaded. Returning to the example in Figure 1, move 1, would of type S, moves 2, 3, 4 and 5 of type E and moves 1^*, 2^*, 3^*, 4^* and 5^* would be of type T.

In principle, during the construction of an ejection chain, we always would like to compute the best possible move when using these rules. However, in order to speed up the procedure, whenever an arc is added in any of these rules, we do not check every possible arc, but restrict us to those arcs (v, w) that seem promising. To this end, we require that either vertex w must be a direct neighbor of vertex v in the *Delaunay triangulation* of the set of customers of the underlying instance[3], or (v, w) is one of the 12 shortest arcs leaving v such that this arc does not imply a time window violation on its own. (No conditions are placed on arcs that involve the depot vertex.) This proved to constitute a good candidate list for our test sets. Let us call a move, i.e., the application of a rule together with the selection of leaving and entering arcs, *admissible* if one of the above conditions for the entering arc is satisfied[4].

2.3 Ejection Chain Construction

The construction of an ejection chain now works as follows. Starting from a given solution, among all admissible S-moves, we select the one that leads to the best possible CDR. Then, we chain a sequence of E-moves from CDR to CDR in the same way, where ejected arcs are declared tabu for the remainder of the ejection chain. The chain is generated until either no more admissible E-moves exist that are not tabu, or a predefined maximum depth of the ejection chain is reached. From every single CDR, we generate all possible trial solutions using admissible T-moves. The *aspiration criterion* of the Tabu Search we use allows previously ejected arcs to be available in T-moves. Eventually, the compound move consists of moving to the best trial solution that has been generated.

In order to be able to move to the 'best' admissible CDR structure for any S- or E-transition, we must decide on the quality of a CDR reference structure (recall that it does not represent a feasible solution). To this end, we associate a simple cost function with the CDR reference structure, by just counting the total length of arcs included in the structure, incremented by a penalty term for the total time window violations of the *implicit routes*. For example, in Figure 5, we can distinguish three implicit routes: the two cycles as seen in Figure 4 and the route starting in c, via t and r back to c.

Finally, it is important to realize that the final algorithm generates several ejection chains after another. Each ejected arc is therefore not only declared tabu for the remainder of the given ejection chain, but all arcs ejected in one ejection chain are declared tabu also for the next θ ejection chains considered by the

---

[3] See [6] for definitions regarding Delaunay triangulations.

[4] Glover uses slightly stricter conditions, so-called *legitimacy conditions*, to determine the arcs that are susceptible to being added or ejected, see [7]. The same legitimacy conditions have also been adopted by Rego [13].

algorithm, where θ is a randomized parameter that is explained in more detail in Section 4. We observed that the procedure performed significantly better if all arcs ejected in the chain were declared tabu, instead of declaring tabu only the ejected arcs involved in the compound move. Apparently the ejected arcs in the remainder of a chain constitute so-called *critical event memory* of the Tabu Search process, see [8].

3   Higher Level Meta-heuristics

Ejection chain procedures are able to manipulate multiple solution components within a single compound move, as explained in the previous section. Based on the theory developed by Glover [7] for the DR reference structure, the CDR-structure proposed in this paper is conjectured to provide a strong form of connectivity between any two solutions. Therefore, a search can (and indeed did) easily get stuck in a so-called *basin of attraction*, see [11]. Hence we need a mechanism that will provide an escape trajectory from such a basin. We considered Simulated Annealing (SA) as well as Iterated Local Search (ILS) to steer the generation of ejection chains. SA requires that neighbor generation and feasibility checking is fast. We experimented with two SA designs, one that returned the best trial solution to an SA process and one that returned all trial solutions to an SA process. In our view, however, the generation of neighbors via ejection chains is too costly for SA to work well. Preliminary testing confirmed our conjecture. In great contrast, the use of ILS in combination with tabu driven ejection chains proved to be very successful. Therefore we restrict further discussions to our experiments using ILS.

3.1   Iterated Local Search

A simple, yet surprisingly effective technique is to apply an *Iterated Local Search* strategy, see [11,2]. We propose an ILS where the embedded heuristic is based on tabu controlled ejection chains and uses random vertex ejections as *kick moves*. From a given start solution S_1 a basic ILS scheme can be stated as follows:

> Apply a kick move on S_1 yielding S_2.
> Perform a Local Search on S_2 resulting in S_3.
> Choose whether or not to accept S_3.
> If S_3 is accepted restart from S_3, otherwise restart from S_1.

We refer to the solution on which the kick move is performed as the *center solution*. We propose the random ejection of a single vertex v as a kick move. The ejected vertex v will form a new route on its own, while we link v's predecessor to its successor. Note that this kick move never introduces infeasibility. One of the main strengths of our procedure is its ability to merge routes in a way similar, but much more efficient, than the well known Clark & Wright savings procedure [3]. Therefore creating single vertex routes presents no problems. In fact, the kick move makes it much easier for our procedure to find new moves, since ejecting vertices always creates more 'freedom' in existing routes. Note that, through the objective, the procedure is strongly biased to decrease the number of routes, so creating a new route by ejecting a vertex through a kick is a move not likely to be made by the procedure otherwise, even though the kick move is available as a forced combination of an $S2$- and $T2$-move. We thus follow the view expressed in [12], namely that a kick move should correspond to a modification of the structure that is not easily accessible to the moves already available or is unlikely to be chosen by the procedure itself. Since we use Tabu Search, instead of strict Local Search, the kick move is temporarily 'irreversible', thus the procedure is forced to find an alternative way to repair the sustained 'damage' obtained from the kick move. Algorithm 1 summarizes our proposed approach.

```
1  Generate Initial Solution
2  while  n < MaxNrOfIterations  do
3       while  Depth < Maximum depth  do
4            Move to best available CDR reference structure
5            Determine best available trial solution
6            Depth++
7       end
8       Determine best trial solution over current ejection chain
9       Update best known solution and center solution
10      n++
11      if  n mod KickFrequency = 0  then
12           Return to center solution
13           Apply kick move
14      end
15      if  No improvement found for α iterations  then
16           Apply diversification move
17      end
18 end
```

Algorithm 1: ILS Ejection Chain Procedure

Ejection chains are generated in lines 3–10. The maximum depth for an ejection chain was 50. Lines 11–14 describe the kick procedure; the kick move was performed every 5 iterations. Frequently performing a low strength kick worked best. Ejecting more than one vertex per kick move proved unsuccessful. Our ILS procedure only accepts improvements. Of course, other acceptance criteria could be used, see [4,11]. The behavior of the ILS procedure is such that, in a sense, we always try to stay close to the current best solution value. This is in line with the so-called *Pyramid Principle*, as stated by Glover [8]. Finally, lines 15–17 provide a diversification measure that takes effect when the ILS itself is unable to improve solutions for a pre-specified number of iterations. To this end, recall that, after each ejection chain, all ejected arcs involved in S- and E-rules are declared tabu for an additional θ ejection chains, where θ is randomly drawn from the interval $[10, 30]$. The ejected arcs involved in the final T-move are also declared tabu. As a diversification measure, we increased this interval to $[70, 100]$ for 100 iterations, if the procedure did not find an improvement for 1000 subsequent iterations. No kick moves are performed during these 100 iterations. The diversification move ends by declaring the best solution in the 100th ejection chain as the new center solution.

4 Computational Results

We present results for two versions of the algorithm, one that minimizes total distance only (DIST), irrespective of the number of vehicles used, and one that also minimizes total distance, but simultaneously penalizes the use of vehicles in the objective function (VEH). The second version was implemented to be able to compare with other heuristics, since they, in contrast to exact methods, primarily focus on minimization of vehicles.

Table 1 shows the computational results on the Solomon VRPTW test instances with 100 clients [15]. For some instances the global minimum distance is known [16], in which case we include the corresponding values in the column labelled 'Global Optima'. The table further shows, per instance, the ARO (discussed below), the results of the two variants of our algorithm, showing respectively the number of vehicles, total distance, and time (in seconds) when the best solution was found, as well as the solutions of another Tabu Search based algorithm (see below). Both versions of the algorithm share the same design, they only differ in the utilized objective function. We used trivial start solutions, in which each client forms an individual route. The algorithms were run for 10 minutes per instance, resulting in an average number of some 300,000 ejection chains per instance. The algorithm is coded in C++, using the Microsoft Visual C++ .NET compiler (2003), on a 2.8Ghz Pentium 4 processor, with 1GB RAM.

The results of Table 1 confirm that the algorithm gets close to the optimal solutions. For instances for which global optimal values are available, the results are on average 0.99% away from the global optimum. These values were found after 170 seconds on average (for the version that minimizes total distance, DIST). When we terminate the same algorithm after 60 seconds, the solutions are 1.67% away from the optimum, on average. Although the method is designed to minimize distance, we were able to obtain reasonable results for minimizing the number of vehicles, too, when we biased the objective function. We compare our results to those of another tabu implementation by Rochat and Taillard [14]. In a review of VRPTW designs in [16] from 2002, the results obtained by this method are referred to as excellent. It can be concluded that the algorithm labelled VEH compares reasonably to Rochat and Taillards results. It results in, on average, half a vehicle more, however with less total distance.

The column labelled ARO displays the average relative overlap (ARO) of any two time windows[5]. We observed that, on average, when the procedure performs worse in terms of solution quality and/or time required to find good solutions, the ARO is high. This makes sense, since a high ARO implies that the instance more closely resembles a VRP without time windows, while our procedure is explicitly designed for settings in which the time windows are tight. In fact, in settings with a lot of overlap between time windows, path reversals might be useful. But as a tribute to time windows, path reversals are explicitly excluded in our algorithm design.

---

[5] $\text{ARO} = 100 \cdot \sum_i \sum_{j>i} \left(\frac{Overlap_{ij}}{|TW_i|} + \frac{Overlap_{ij}}{|TW_j|} \right) / (n(n-1))$, where $|TW_i|$ is the length of time window i and n is the number of clients (the depot is not considered a client).

Table 1. Results on the Solomon VRPTW instances with 100 clients

Solomon 100		EC MIN DIST			EC MIN VEH			R-T [14]		Global Optima	
SET	ARO	VEH	DIST	TIME	VEH	DIST	TIME	VEH	DIST	VEH	DIST
C101	6	10	828.9	1.2	10	828.9	0.3	10	828.9	10	827.3
C102	29	10	828.9	9.5	10	828.9	41	10	828.9	10	827.3
C103	53	10	828.1	28.7	10	828.1	312.5	10	828.1	10	826.3
C104	76	10	829.0	378	10	840.0	390.8	10	841.6	10	822.9
C105	12	10	828.9	6	10	828.9	10.2	10	828.9	10	827.3
C106	15	10	828.9	0.9	10	828.9	7.9	10	828.9	10	827.3
C107	18	10	828.9	62.8	10	828.9	18.3	10	828.9	10	827.3
C108	25	10	828.9	5.6	10	828.9	2.1	10	828.9	10	827.3
C109	38	10	828.9	58	10	828.9	10.7	10	828.9	10	827.3
C201	4	3	591.6	2.9	3	591.6	1.4	3	591.6	3	589.1
C202	28	3	591.6	19	3	591.6	4.5	3	591.6	3	589.1
C203	52	3	591.2	11.4	3	591.2	57.5	3	591.2	3	588.7
C204	76	3	590.6	308.2	3	594.5	594.4	3	597.8	-	-
C205	94	3	588.9	4.6	3	588.9	7	3	588.9	3	586.4
C206	15	3	588.5	8.1	3	588.5	3.1	3	588.5	3	586.0
C207	19	3	588.3	20.1	3	588.3	13.3	3	588.5	3	585.8
C208	20	3	588.3	25.6	3	588.3	7.2	3	588.5	3	585.8
R101	6	20	1642.9	34.4	19	1650.8	38	19	1656.2	20	1637.7
R102	28	18	1473.7	268.6	18	1473.7	420.7	18	1477.4	18	1466.6
R103	51	15	1227.8	320.6	14	1218.7	319.3	14	1222.9	14	1208.7
R104	74	11	1006.6	169.4	10	1007.8	432.2	10	1013.3	-	-
R105	19	15	1365.5	132.5	15	1367.5	114.2	14	1404.8	15	1355.3
R106	36	13	1246.1	514.9	13	1243.9	219.8	12	1293.9	13	1234.6
R107	56	12	1097.8	240.5	11	1092.0	24.8	11	1085.8	11	1064.6
R108	77	10	969.2	286	10	980.2	244.1	10	965.3	-	-
R109	38	13	1161.8	440.3	12	1172.6	253.6	12	1186.4	13	1146.9
R110	56	12	1108.2	493.3	12	1102.4	284.2	11	1107.9	12	1068.0
R111	55	12	1088.7	203.4	11	1092.2	47	11	1070.9	12	1048.7
R112	78	11	981.7	583.9	10	1000.2	314.5	10	965.7	-	-
R201	13	8	1150.0	390.7	6	1171.8	382.3	4	1485.4	8	1143.2
R202	34	7	1041.2	468.1	6	1049.5	599.4	4	1101.5	-	-
R203	55	6	877.6	182	5	890.0	437.7	4	913.0	-	-
R204	77	5	751.2	598.5	3	808.8	103.8	3	824.6	-	-
R205	28	6	966.7	216.6	4	1009.5	18.9	3	1205.6	-	-
R206	44	5	902.1	198.5	5	896.4	433.9	3	956.1	-	-
R207	62	4	813.1	387.4	3	856.0	472.7	3	814.8	-	-
R208	81	4	725.7	215.6	3	710.5	306.8	3	708.8	-	-
R209	41	5	870.5	572.8	5	881.4	346	4	901.9	-	-
R210	44	7	927.0	473.9	4	931.6	486.7	3	1087.3	-	-
R211	59	5	779.3	530.3	4	770.7	320.4	3	794.5	-	-
RC101	19	16	1645.6	20.2	15	1663.4	153.2	15	1737.0	15	1619.8
RC102	36	15	1487.6	532	14	1505.1	351.7	13	1480.7	14	1457.4
RC103	55	12	1289.1	206.1	13	1360.7	599.6	11	1264.3	11	1258.0
RC104	76	11	1187.0	142.2	11	1169.1	192.4	10	1157.2	-	-
RC105	36	16	1544.0	552.1	15	1562.7	274.7	15	1543.2	15	1513.7
RC106	40	14	1412.7	339.1	13	1404.4	428	12	1415.6	-	-
RC107	59	12	1235.2	580.8	11	1255.1	414.3	11	1262.4	-	-
RC108	75	12	1145.9	52.1	11	1162.9	259.6	11	1149.6	-	-
RC201	14	9	1268.8	285.7	5	1327.2	162.3	5	1469.7	9	1261.8
RC202	34	8	1113.6	197.1	5	1160.6	281.2	4	1443.7	-	-
RC203	56	6	955.4	146.3	4	1015.2	170.5	4	1014.0	-	-
RC204	77	4	791.4	207.2	4	812.1	438	3	843.1	-	-
RC205	28	7	1167.6	128	5	1290.0	144.5	5	1286.7	-	-
RC206	30	6	1080.1	418.1	5	1067.0	515	4	1207.8	-	-
RC207	44	7	988.2	74.5	5	996.5	589	4	1079.1	-	-
RC208	63	6	817.5	449	5	789.5	380.5	3	919.8	-	-
Averages		8.9	990.8	235.8	8.2	1002.0	240.3	7.7	1030.6	-	-

5   Conclusions and Recommendations

One of the key reasons for the good performance of our algorithms for VRPTW is, we believe, due to the ability to generate powerful compound moves that do not require a path reversal. It must be noted, however, that methods that do perform path reversals, are extremely efficient in settings without time windows. We observed that, on average, when there is high overlap in the time windows, the procedure performs less effective. It is likely that the absence of path reversals is a reason. However, our concept can be adapted quite easily to include moves that use path reversals, too.

It can be considered a strength that the procedures do not, in any form, use pre-processing or post-processing. Also, the procedures cannot be considered two-phased methods, since they always use the same, extremely simple, start solution. A strong feature of the Iterated Local Search is that the kick move can be made very problem-specific, and can be used to decrease any possibly sustained infeasibility during the search.

Generalizing the CDR reference structure by relaxing the constraint that one of the roots must be the core is likely to further improve the procedure. The resulting *Generalized Doubly Rooted* reference structure (GDR) is shown in figure 6. The stated rules can be used as guidelines to create new S, E and T-type rules to exploit the GDR structure. E rules can be constructed that are able to increase or decrease the number of routes. In contrast, in the stated procedure, the total number of routes cannot be changed by more than one route per ejection chain. Further research is necessary to examine the potential of the GDR structure over the CDR structure. Finally, the procedure can, in

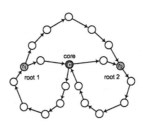

Fig. 6. Generalized Doubly Rooted reference structure (GDR)

all probability, easily be extended to the multi-depot case by using the reference structure in Figure 6, but allowing the roots to lie on cycles not necessarily joined by the same depot.

Acknowledgements. We thank Geert Teeuwen for the cooperation at CQM. In addition, the authors would like to thank Fred Glover and Emile Aarts for all their valuable comments, suggestions and encouragement. We found their collaboration very inspiring. Finally, we thank the anonymous referees for their valuable suggestions for improvement.

References

1. E. H. L. Aarts and J. K. Lenstra. *Local search in combinatorial optimization.* Wiley, Chichester, UK, 1996.
2. C. Blum and A. Roli. Metaheuristics in combinatorial optimization: Overview and conceptual comparison. *ACM Comput. Surv.*, 35(3):268–308, 2003.
3. G. Clark and J. W. Wright. Scheduling of vehicles from a central depot to a number of delivery points. *Operations Research*, 12:568–581, 1964.
4. M. den Besten, T. Stützle, and M. Dorigo. Design of iterated local search algorithms. In E. J. W. Boers, J. Gottlieb, P. L. Lanzi, R. E. Smith, S. Cagnoni, E. Hart, G. R. Raidl, and H. Tijink, editors, *Applications of Evolutionary Computing*, volume 2037 of *Lecture Notes in Computer Science*, pages 441–451, 2001.
5. D. Gamboa, C. Rego, and F. Glover. Implementation analysis of efficient heuristic algorithms for the traveling salesman problem. *Computers and Operations Research*, 2005. To appear.
6. P. L. George and H. Borouchaki. *Delaunay Triangulation and Meshing, Applications to Finite Elements.* Hermes, 1998.
7. F. Glover. Ejection chains, reference structures and alternating path methods for traveling salesman problems. *Discrete Applied Mathematics*, 65:223–253, 1996.
8. F. Glover and M. Laguna. *Tabu Search.* Kluwer, Dordrecht, NL, 1998.
9. P. Kilby, P. Prosser, and P. Shaw. Guided local search for the vehicle routing problem. In S. Voss, S. Martello, I. H. Osman, and C. Roucairol, editors, *Meta-Heuristics: Advances and Trends in Local Search Paradigms for Optimization*, pages 473–486. Kluwer, Boston, MA, 1997.
10. S. Lin and B. W. Kernighan. An effective heuristic for the traveling salesman problem. *Operations Research*, 21:498–516, 1973.
11. H. R. Lourenco, O. Martin, and T. Stützle. Iterated local search. In F. Glover and G. Kochenberger, editors, *The Handbook of Metaheuristics*, pages 321–353. Kluwer, Norwell, MA, 2002.
12. O. C. Martin and S. W. Otto. *Combining Simulated Annealing with Local Search Heuristics*, volume 63 of *Annals of Operations Research*, pages 57–75. 1996.
13. C. Rego. A subpath ejection method for the vehicle routing problem. *Management Science*, 44(10):1447–1459, 1998.
14. Y. Rochat and É. D. Taillard. Probalistic diversification and intensification in local search for vehicle routing. *Journal of Heuristics*, 1:147–167, 1995.
15. M. Solomon. Algorithms for the vehicle routing and scheduling problem with time window constraints. *Operations Research*, 35:254–265, 1987.
16. P. Toth and D. Vigo. *The Vehicle Routing Problem.* Society for Industrial and Applied Mathematics, Philadelphia, PA, 2002.
17. J. Xu and J. Kelly. A network-flow based tabu search heuristic for the vehicle routing problem. *Transportation Science*, 30:379–393, 1996.

3D Inter-subject Medical Image Registration by Scatter Search[*][**]

Oscar Cordón[1], Sergio Damas[2], J. Santamaría[2], and Rafael Martí[3]

[1] Dept. of Computer Science and A.I., University of Granada,
ocordon@decsai.ugr.es
[2] Dept. of Software Engineering, University of Granada
sdamas@ugr.es, jsantam@ugr.es
[3] Dept. of Statistics and Operational Research, University of Valencia
rafael.marti@uv.es

Abstract. Image registration is a very active research area in computer vision, namely it is used to find a transformation between two images taken under different conditions. Point matching is an image registration approach based on searching for the right pairing of points between the two images. From this matching, the registration transformation we are searching, can be inferred by means of numerical methods.

In this paper, we propose a scatter search (SS) algorithm to solve the matching problem. SS is a hybrid metaheuristic with a good trade-off between search space diversification and intensification. On the one hand, diversity is basically introduced from a population-based approach where systematic combinations of subsets of solutions are performed. On the other hand, intensification is achieved with a local search procedure, to ensure the local improvement of promising solutions. Our computational experimentation in a real-world inter-subject medical registration environment establishes the effectiveness of our procedure in relation to different approaches usually applied to solve the problem.

1 Introduction

Image registration (IR) is a fundamental task in computer vision used to finding a correspondence (or transformation) among two or more pictures taken under different conditions: at different times, using different sensors, from different viewpoints, or a combination of them [2]. Point matching is an IR approach based on searching for the right pairing of points between two images. From this matching, the registration transformation we are searching, can be inferred by means of numerical methods. Therefore, the hard problem to be faced from this IR approach is point matching.

[*] Research by Oscar Cordón and Sergio Damas is supported by the Ministerio de Ciencia y Tecnologia under project TIC2003-00877 (including FEDER fundings).
[**] Research by Rafael Martí is partially supported by the Ministerio de Educación y Ciencia (refs. TIN2004-20061-E and TIC2003-C05-01) and by the Agencia Valenciana de Ciencia i Tecnologia (ref. GRUPOS03 /189).

M.J. Blesa et al. (Eds.): HM 2005, LNCS 3636, pp. 90–103, 2005.

In the last few years, there is an increasing interest on applying metaheuristic approaches to IR [4,5,6,13,17,22]. In this work, we try to exploit the benefits of applying *Scatter Search* (SS) [15] to solve a real-world problem of inter-subject medical registration of human brains and our contributions are related to the fact of properly balancing the diversity and intensification components of this hybrid metaheuristic as well as taking advantage of heuristic information extracted from the images (curvature information) to guide the process. Such information has been proved to return good quality solutions [7]. We consider it in order to reduce the number of meaningful points and to obtain a good topological characterization of the shape, therefore enabling a better matching. The proposed SS-based IR algorithm is a significantly improved version of that in [4] that allows us to properly deal with more complex IR problems (see Section 7).

The performance of the SS algorithm proposed is tested against the preliminary version [4] in order to demonstrate the improvement made. Besides, it is compared to a simple greedy algorithm, as well as to the technique usually applied in the computer vision field, Iterative Closest Point (ICP) [1,7,16]. Finally, we have also considered a recent hybrid approach combining ICP with a Simulated Annealing procedure (ICP+SA) [17].

To do so, in Section 2 we first briefly describe the IR problem. Section 3 is devoted to introduce the SS methodology. The process to extract the heuristic information from the image used to guide the search in the solution space is presented in Section 4. Next, we describe the objective function in Section 5, and our implementation to solve the 3D IR problem in Section 6. Finally, the paper ends with computational experiments (Section 7), concluding remarks and open research lines (Section 8).

2   Image Registration

IR is a difficult optimization problem and can be simply stated as finding a mapping between two images: I_1 named *scene*, and I_2 named *model*. The objective is to find the mathematical transformation f that applied to I_1 obtains I_2. Generally speaking, an image is stored in a huge amount of pixels, therefore most IR methods usually apply a preprocessing to extract the most relevant geometric primitives (point, lines, etc) that, in a certain way, define the objects contained in the image. Therefore, in these *feature-based* methods, the problem is reduced to find the transformation between two sets of geometric primitives. In this paper, we restrict our attention to the case of two sets of primitives P_1 and P_2, consisting uniquely of points ($P_1 \subseteq I_1$, $P_2 \subseteq I_2$).

Although the final registration problem solution consists of the right values for the parameters which determine f, two different approaches arise, each of them working in a different solution space: i) to search for the optimal point matching between the two images and then identify the appropriate transformation parameters —using numerical methods such as least square estimation— to overlay the scene and the model considering such matching ([1,6,7,8,16]); and ii) to directly search in the parameter space (usually by means of evolutionary algorithms) to get the estimated transformation ([5,13,14,22]). While the former involves determining which of the scene points matches each model one (thus becoming an NP-hard combinatorial optimization problem) and then indirectly computing the transformation parameters; the latter deals with the direct estimation of the registration transformation causing this model-scene overlapping.

The point-matching approach is probably the most classical method in feature-based registration. In mathematical terms, it can be described as follows. Given two set of points $P_1 = \{x_1, x_2, ..., x_n\}$ and $P_2 = \{y_1, y_2, ..., y_m\}$, the problem is to find a transformation f such that $y_i = f(x_{\pi(i)})$ for $i = 1, ..., r$ ($r = min(n, m)$), where π is a permutation of size l (with l being the maximum between n and m). Without loss of generality and to simplify the notation, we consider that P_1 is the larger point set, i.e., its dimension n is greater than that of P_2, m.

The problem solving is naturally divided in two phases. In the first one, a permutation of l elements defines the matching between the points in P_1 and P_2 in such a way that the first r elements ($r = m$ in our case) of π are the P_1 points associated to each of the m P_2 points. In the second phase, from this matching of points and using a numerical optimization method (usually least squares estimation), the parameters defining the transformation f are computed. The goal is to find the transformation minimizing the distances between the model points and the corresponding transformed scene points. Therefore, in optimization terms, the value associated with permutation π is given by the expression:

$$g(\pi) = \frac{\sum_{i=1}^{r} \| f_\pi(x_{\pi(i)}) - y_i \|^2}{r},$$

i.e., $g(\pi)$ corresponds to the *Mean Square Error (MSE)*. Therefore, the point matching problem can be simply stated as minimizing $g(\pi)$ for any permutation π of l elements and its corresponding transformation f. In this paper, we face the IR problem from this point matching approach, proposing a SS implementation to find high quality solutions to this combinatorial optimization problem.

3   Fundamentals of Scatter Search

SS was first introduced in [9] as a heuristic for integer programming. The key idea of SS is the exploration of the solution space operating on a set of solutions, the reference set, by combining these solutions systematically to create new ones. When the main mechanism for combining solutions is such that a new solution is created from the linear combination of two other solutions, the reference set may evolve. SS does not leave solutions in a raw form produced by its combination mechanism, but subjects candidates for entry into the reference set to heuristic improvement, as we elaborate subsequently.

The reference set of solutions in SS is relatively small. SS chooses two or more elements of the reference set in a systematic way with the purpose of creating new solutions. Since the number of two-element to five-element subsets of a reference set, for example, can be quite large, even a highly selective process for isolating

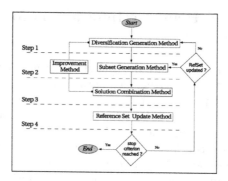

Fig. 1. The SS control diagram

preferred instances of these subsets as a basis for generating combined solutions can produce a significant number of combinations, and so there is a practical need for keeping the cardinality of the set small. Typically, the reference set has 20 solutions or less. SS systematically injects diversity to the reference set.

The fact that the mechanisms within SS are not restricted to a single uniform design allows the exploration of strategic possibilities that may prove effective in a particular implementation. Different authors ([10,11,15]) have recently proposed a template for implementing SS (see Figure 1):

1. A *Diversification Generation Method* to generate a collection of diverse trial solutions, using an arbitrary trial solution (or seed solution) as an input.
2. An *Improvement Method* to transform a trial solution into one or more enhanced trial solutions. Neither the input nor the output solutions are required to be feasible, though the output solutions will more usually be expected to be so. If no improvement of the input trial solution results, the "enhanced" solution is considered to be the same as the input solution.
3. A *Reference Set Update Method* to build and maintain a reference set consisting of the b "best" solutions found (as said, b is typically no more than 20), organized to provide efficient accessing by other parts of the method. Solutions gain membership to the reference set according to their quality or their diversity.
4. A *Subset Generation Method* to operate on the reference set, to produce a subset of its solutions as a basis for creating combined solutions.
5. A *Solution Combination Method* to transform a given subset of solutions produced by the Subset Generation Method into one or more combined solution vectors.

Of the five methods in the SS methodology, only four are strictly required. The Improvement Method is usually needed if high quality outcomes are desired, thus making the algorithm become a hybrid metaheuristic, more specifically, a kind of memetic algorithm [19]. On the other hand, more advanced hybrid SS designs could incorporate a short term tabu search or other complex metaheuristic as the improvement method instead of classical local search (usually demanding more running time).

4   Shape-Derived Heuristic Information for 3D Image Registration

This section is devoted to introduce the heuristic information that can be derived from the curvature of the shapes included in the images to be registered in order to better solve the IR problem. To do so, let us first define the iso-intensity surface of a 3D image, which will be called simply the iso-surface in the rest of this paper. For any continuous function $C(x, y, z)$ of $\mathbb{R}^3$, any value I of $\mathbb{R}$ (called the iso-value) defines a continuous, not self-intersecting surface, without hole, which is called the iso-intensity surface of C [18]. A non ambiguous way to define the iso-surface is to consider it as being the surface which separates the space regions where the intensity of C is greater or equal to I from these regions whose intensity is strictly lower than I. Whether such an iso-surface corresponds to the boundary of the scanned object is another problem, that will not be considered in the current contribution. Because of their good topological properties, iso-surface techniques are the most widely used methods of segmentation for 3D medical images.

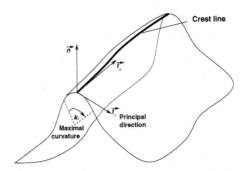

Fig. 2. Differential characteristics of surfaces

Let us see now some properties of the iso-surfaces (see Figure 2). At each point p of those surfaces, there is an infinite number of curvatures but, for each direction t in the tangent plane at p, there is only one associated curvature k_t. There are two privileged directions of the surface, called the principal directions (t_1 and t_2), which correspond to the two extremal values of the curvature: k_1 and k_2. There are many more parameters to determine the characterization of surfaces, but we have considered the latter two (k_1 and k_2) being enough to be considered as heuristic information to help us the IR problem solving.

5   Objective Function

As said, the main novelty of our method is that the features of the image iso-surfaces (the curvature information seen in Section 4) are used to guide the matching. So, we define a function $m_{error}(\cdot)$ evaluating the goodness of the matching stored in π by using the curvature values shown in Section 4. In this contribution, we have chosen the following:

$$m_{error}(\pi) = \Delta k_1 + \Delta k_2 \quad \text{where} \quad \Delta k_j = \sum_{i=1}^{r}(k_j^i - k_j^{\pi_i})^2, \; j = \{1,2\}$$

Δk_1 and Δk_2 measure the error associated to the matching of scene and model points with different values for the first and second principal curvatures, respectively.

This way, the objective function (noted $F(\cdot)$) will include information regarding the usual IR measure $g(\pi)$ (see Section 2) and the previous criterion as follows:

$$\min F(\pi) = w_1 \cdot g(\pi) + w_2 \cdot m_{error}(\pi) \qquad (1)$$

where the first term stands for the transformation estimation error and the second one for the matching error, and w_1, w_2 are weighting coefficients defining the relative importance of each.

With such a function we will have a more suitable similarity measurement to make a better search process in the space of solutions. Instead of considering a function based on a single registration error criterion, the use of the previous two terms working together to solve the IR problem is an important part of our novel proposal.

6   A Scatter Search Algorithm for 3D IR

As seen in Section 3, the SS methodology basically consists of five elements (and their associated strategies): Diversification Generation Method, Improvement Method, Solution Combination Method, Reference Set Update Method, and Subset Generation Method. The three former are problem dependent, and should be designed specifically for the problem at hand (although it is possible to design "generic" procedures, it is more effective to base the design on specific characteristics of the problem setting). The other two are context independent, and usually have a standard implementation.

In this contribution, we have considered the same design of the preliminary version [4] for the two context independent components, the Reference Set Update Method, and the Subset Generation Method. However, the implementation of two of the three specific elements (the Diversification Generation Method and the Improvement Method) has been changed in order to improve the performance of our SS-based IR technique, allowing it to deal with significantly more complex problem instances as we will see in the next section. Another design choice has been also considered for the last component, the Solution Combination Method, but the experimental results showed that the original operator behaves better.

The next five items are devoted to describe the design decisions made for each SS component in our problem, especially focusing on the description of the two specific SS elements mentioned above: the Diversification Generation Method, and the Improvement Method, for which a wider description is reported.

Diversification Generation Method. As seen, instead of considering a completely random generation of the initial solutions as done in [4], a heuristic procedure can be designed by means of the image-derived information described in Section 4. It can be used to establish a preference for good assignments between the scene image points and the model image ones. Hence, a point x_i from the scene image is more likely to be assigned to those model points y_j presenting the same or similar curvature values k_1 and k_2.

Hence, we can make use of this information in order to generate the initial set P of diverse solutions for our SS procedure, thus obtaining solutions with both good quality and high diversity. To do so, instead of fixing a selection order for the scene points x_i and then assigning the closest model point y_j (as regards the curvature values) not yet considered to each of them (which would result in a deterministic, greedy heuristic), we introduce randomness in both processes allowing each decision to be randomly taken among the best candidates. This way, our diversification generation method behaves similarly to a GRASP construction phase [20].

The reference set, *RefSet*, is a collection of $b = b_1 + b_2$ solutions (reference points) that are used to generate new solutions. The construction of the initial reference set starts with the selection of the best $b_1 \leq b$ improved solutions from P. These solutions are added to *RefSet* and deleted from P. The remaining $b_2 = b - b_1$ *RefSet* solutions are selected from P taking into account the diversity.

Improvement Method. It involves applying local search strategies. In our work, we have chosen the "best-first" local search with the swapping neighbor operator in which swaps are used as the primary mechanism to move from one solution to another. In our previous work [4], two improvements were considered in order to speed the local search procedure up. On the one hand, a primary strategy was applied in the neighborhood generation by only considering promising swapping moves taking as a base the curvature information. On the other hand, a selective application of the local optimizer was also considered. As mentioned in Section 3, a SS algorithm can be implemented without this component, although its use allows it to obtain high quality outcomes. Hence, in order to obtain an appropriate balance between the solution quality upgrade resulting from the Improvement Method use and the time consumed by it, we decided not to run the local search over each solution generated by the Combination Method but only on some of them.

Note that the contribution to the objective function F of every matching stored in π is different. Therefore, points should not be treated equally by a procedure that selects an index for a local search (i.e., for search intensification). Hence, as a more suitable extension of our previous proposal, we are now introducing the sorting of the point-matchings considering the contribution of each correspondence to the permutation evaluation. Those matchings with a high error contribution will be good candidates for a swapping. In order to look for a new correspondence for such points we will use the curvature information to favor a matching between similar feature points.

Subset Generation Method. In our implementation, subsets are always composed of pairs of solutions and three different kinds of them are generated. On the one hand, subsets with the $b_1 * (b_1 - 1)$ possible pairs of solutions in the quality part of the *RefSet* are created in order to intensify the search by combining high quality solutions. On the other hand, each of the $b_2 * (b_2 - 1)$ pairs of solutions in the diversity part are also considered to generate combined solutions for diversification purposes. Finally, a third group of $b_1 * b_2$ subsets is created by pairing each solution of the quality part with every one in the diversity part, thus getting combined solutions with an intermediate search behavior.

Solution Combination Method. This method takes each solution subset built by the *Subset Generation Method* and obtains new solutions from it by applying a combination operator. As our coding scheme is based on a permutation, in [4], we made use of the classical *PMX* recombination operator [12] to combine the two solutions from each subset. We have now tested another type of combination method, named *Voting Method* [3]. It is based on deterministic elements and is widely used in the context of applying SS to tackle permutation-based problems [15]. Unfortunately, the results worsened using the latter approach. Therefore, we decided to maintain the use of the PMX combination method.

Reference Set Update Method. In our approach, the *Solution Combination Method* first generates all the new solutions from the combinations of each subset, and then the reference set is updated with these new solutions. Such updating process is called a *static updating mechanism* [15].

The method makes an updating process using both quality and diversity criteria. Firstly, the list of new solutions is increasingly sorted by their fitness value: $\{s_1 \leq s_2 \leq \ldots \leq s_{(b^2-b)/2}\}$ where the first solution s_1 is the best one generated by the combination method in the current iteration. If the fitness value of s_1 is lower than the worst solution in the quality part of *RefSet* ($\{q_1 \leq q_2 \leq \ldots \leq q_{b_1}\}$), i.e. lower than q_{b_1}, then s_1 will replace q_{b_1}. The process goes on checking if the fitness value of s_2 is lower than the one of q_{b_1-1}, and so on.

After considering the quality criterion, if the list is not empty the diversity updating will start. To do so, we will analyze the diversity of every solution in the list compared to those in *RefSet*, in order to guarantee that the b_2 most diverse ones are stored in the diversity part of RefSet. The process goes on with the remaining solutions in the list till every solution in the list is checked.

Finally, in case no solution in the Reference set is updated, a restart process is applied. The process involves maintaining the best solution from the quality part of *RefSet*, generating the remaining solutions by the usual *Diversification Generation Method*, and starting again the SS procedure.

7   Experiments

7.1   Image Registration Problems Considered

Our results correspond to a number of registration problems for three different 3D images. These images have been obtained from the *BrainWeb* database at

McGill University. The purpose of this simulator is to provide researchers with ground truth data for image analysis techniques and algorithms. *BrainWeb* has been widely used by the IR research community (see, for example, [21]). This experimental setup is significantly more complex than that considered in [4] due to two main reasons. On the one hand, one of the two images handled in the latter was a simple, synthetic one, while in the current experimentation we deal with three real-world magnetic resonance images (MRIs) of three human brains. On the other hand, the most important challenge associated to the current experimentation is that the goal of the IR process is to register *two different images of similar objects*, thus having a more realistic situation in medical IR named *inter-subject registration*. Of course, the registration of different objects is much more complex than that of different views of the same objects, and this was what motivated the extension of our previous algorithm in order to obtain good performance in the new IR scenario. Finally, we should note that four different registration transformations are considered (while only three were used in [4]), thus having two more problem instances (eight instead of six).

These MRIs have suffered the same four global similarity transformations T_i (see Table 1) thus generating the eight different IR problem instances (I_1 vs. $T_i(I_2)$ and I_1 vs. $T_i(I_3)$, $i = 1, \ldots, 4$) to be solved by the old version of the SS-based IR process [4], the new version proposed in this paper, the greedy algorithm described in the *Diversification Generation Method* in Section 6, as well as by the last version of the technique usually applied in the computer vision field, the Iterative Closest Point (ICP) [16]. We have also considered a recent hybrid approach combining ICP with a Simulated Annealing procedure (ICP+SA) [17]. Notice that we consider several more comparison algorithms than in [4] where only an old version of the ICP method and a memetic algorithm designed by ourselves were used.

Table 1. Transformations applied to every 3D image

	T_1	T_2	T_3	T_4
$RotAngle°$	115.0	168.0	235.0	276.9
$RotAxis_x$	-0.863868	0.676716	-0.303046	0.872872
$RotAxis_y$	0.259161	-0.290021	-0.808122	0.436436
$RotAxis_z$	0.431934	0.676716	0.505076	-0.218218
Δ_x	-26.0	6.0	16.0	12.0
Δ_y	15.5	5.5	-5.5	5.5
Δ_z	-4.6	-4.6	-4.6	-24.6
S	1.0	0.8	1.0	1.2

After preprocessing the three images (I_1, I_2, I_3), 583, 393 and 348 points are obtained, respectively (see Figure 3).

7.2 Parameter Settings

All the runs have been performed on a 2.6 GHz. Intel Pentium IV processor. As regards the ICP parameter values, a maximum number of 40 iterations is

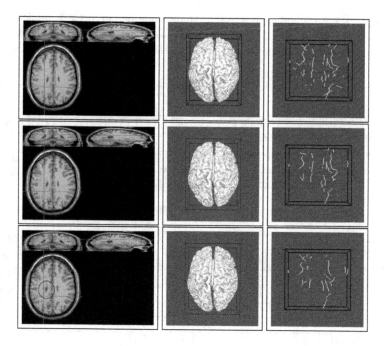

Fig. 3. From left to right, and top to bottom: Original images, their respective isosurfaces, and their crest lines points. Note that the second and third MRIs include 1% of Gaussian noise. I_3 (last row) also considers a multiple sclerosis lesion (see circle).

considered in order to allow a proper convergence of the algorithm. We have also followed the author guidelines to fix the k parameter [16]. The hybrid algorithm (ICP+SA) also considers a maximum number of 40 iterations for ICP and 20 iterations for the annealing, as well as 50 trial movements around each annealing iteration. Meanwhile, both SS variants are run for 20 seconds in every problem instance. Besides, we consider an initial set P with 80 solutions, a number of $b = b_1 + b_2 = 10$ solutions that compound the RefSet, with $b_1 = 7$ and $b_2 = 3$. For the Improvement Method, a maximum of 80 iterations has been fixed at each execution. We perform 15 runs of ICP+SA and the two SS algorithms for every problem instance, starting from a different random number generator seed, in order to avoid the bias of randomness.

7.3   Results

We present the g (i.e. the *MSE*) value (see Section 2) of every solution as a metric to comparing the greedy, ICP, ICP+SA and the two variants of the SS algorithm. Such statistical values are presented in Table 2 where SS stands for the old version of the SS-based IR process [4] and SS* for the improved version presented in this paper.

Regarding to these results, we want to remark that the new version of our SS outperforms the remaining algorithms in every problem instance (consider-

Table 2. MSE corresponding to the four transformations in Table 1 applied to both the I_1 vs. $T_i(I_2)$ and I_1 vs. $T_i(I_3)$ IR problems considered. m stands for the minimum, M for the maximum, μ for the mean, and σ for the standard deviation values.

I_1 **vs.** $T_i(I_2)$										
I_1 **vs.** $T_1(I_2)$						I_1 **vs.** $T_2(I_2)$				
	Greedy	ICP	ICP+SA	SS	SS*		Greedy	ICP	ICP+SA	SS	SS*
m	50.6	344.1	247.3	76.6	**32.5**	m	40.2	131.4	131.4	65.4	**38.4**
M	-	-	344.1	770.1	**38.5**	M	-	-	131.4	112.8	**47.4**
μ	-	-	307.1	315.3	**35.6**	μ	-	-	131.4	90.9	**43.1**
σ	-	-	37.8	209.8	**1.9**	σ	-	-	0	18	**2.2**

I_1 **vs.** $T_3(I_2)$						I_1 **vs.** $T_4(I_2)$				
	Greedy	ICP	ICP+SA	SS	SS*		Greedy	ICP	ICP+SA	SS	SS*
m	63.6	893.6	456.8	165.9	**58.1**	m	72.8	632	283.4	93.4	**50.4**
M	-	-	711	922.1	**63.3**	M	-	-	610.6	629.4	**59**
μ	-	-	558.9	313.4	**60.3**	μ	-	-	465.1	276.5	**53.8**
σ	-	-	80.8	196.2	**1.3**	σ	-	-	101.4	123.5	**2.6**

I_1 **vs.** $T_i(I_3)$										
I_1 **vs.** $T_1(I_3)$						I_1 **vs.** $T_2(I_3)$				
	Greedy	ICP	ICP+SA	SS	SS*		Greedy	ICP	ICP+SA	SS	SS*
m	235.8	517.6	304.5	148.3	**70.3**	m	156.2	329.8	236.5	85.5	**39.8**
M	-	-	432	270.3	**142.6**	M	-	-	297	179.8	**91.8**
μ	-	-	342.7	215.1	**96.1**	μ	-	-	260.9	133.8	**55.1**
σ	-	-	31.8	39.6	**19.7**	σ	-	-	18.1	29.2	**15.1**

I_1 **vs.** $T_3(I_3)$						I_1 **vs.** $T_4(I_3)$				
	Greedy	ICP	ICP+SA	SS	SS*		Greedy	ICP	ICP+SA	SS	SS*
m	236.2	438.4	278.5	99.5	**45.1**	m	334.1	478.4	336.4	114.1	**112.3**
M	-	-	389.2	275.6	**134.3**	M	-	-	428.9	382.8	**143.2**
μ	-	-	346.6	199.4	**61.8**	μ	-	-	381.6	288.4	**122.6**
σ	-	-	33.1	52.7	**23.6**	σ	-	-	24.1	80.8	**8.2**

ing any pair of transformations and images). Moreover, this behavior does not depend on a given run of the SS. As said, fifteen different seeds have been considered for the fifteen runs of this method. If we analyze the MSE mean values, as well as their corresponding standard deviations, two different conclusions arise. First, the MSE mean values associated to SS* are better than the best result achieved by any other algorithm in every problem instance but I_1 vs. $T_2(I_2)$, where the greedy result is slightly better than our SS* mean value (but not than the SS* best one). The good performance of the simple greedy algorithm in this problem instance is due to the heuristic information considered that allows it

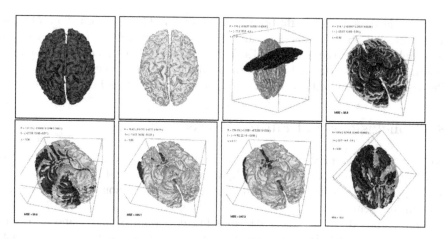

Fig. 4. From left to right, and top to bottom: Isosurface images corresponding to the brain of the I_1 and I_2 MRIs in Figure 3. The third image in the first row show the initial scenario of the transformation T_1 to be estimated by every method. The fourth image in this row corresponds to the best $T1$ estimation (related to our present SS-based IR proposal SS*). The second row respectively shows the estimation of the greedy algorithm, ICP, ICP+SA and the previous SS-based approach [4]. Note that, the higher the overlapping between dark and light grey, the better the $T1$ estimation.

to perform properly in the first IR scenario (I_1 vs. $T_i(I_2)$) where I_1 and I_2 are similar enough to compose a simpler scenario than the I_1 vs. $T_i(I_3)$ one. This can also be appreciated in view of the poor results of the old SS version which does not make a suitable use of that image-derived information. On the other hand, the low standard deviations show the robustness of SS*. In particular, we achieve one of our goals respect to our previous work [4], that is, to design a competitive IR method even in a complex scenario of inter-subject medical IR.

Figure 4 graphically depicts the estimation of the transformation involved in the I_1 vs. $T_1(I_2)$ problem instance, achieved by every algorithm. The first row corresponds to the initial scenario with I_1 (in dark grey, first image), I_2 (in light grey, second image) and both I_1 and $T_1(I_2)$ (third image in the first row).

All of the images in Figure 4 correspond to a preprocessed MRI (see second column of Figure 3). Instead of showing the crest line points extracted at the final step of the preprocessing, we prefer to illustrate the algorithms behavior using the previous step (isosurface extraction). The reason is to ease the geometrical comprehension of the transformation estimated by every algorithm (although, as we mentioned above, every method only works on a selected set of points).

Every algorithm estimates T_1 returning $T_1^{Greedy}, T_1^{ICP}, T_1^{ICP+SA}, T_1^{SS}$ and $T_1^{SS^*}$. Every estimation has been applied to the scene image in order to know its goodness respect to the optimal transformation T_1. This is what the last picture in the first row and those in the second row of Figure 4 show, how close is the estimation of every algorithm respect to T_1. The higher the overlapping between the dark and the light brains, the better the T_1^i estimation. None of these algorithms but SS* achieves the near optimal transformation T_1 (final goal of the IR problem). From the results of Table 2, we can see that this is not an isolated instance, but every algorithm maintains this behavior in every T_i estimation. Because of space limitation, we can not show the corresponding figures. ç

8   Conclusions and Future Works

We have presented an SS-based IR method involving on an extension of a previous proposal [4]. We have obtained significant results in real inter-subject medical experiments, clearly outperforming the previous version as well as several existing IR techniques.

Our SS-based IR process is still in a full development state, and there are many new possible designs for each of the SS components to achieve more accurate solutions than those obtained in this first study. One of such new extensions and alternative SS method designs can be that making a more aggressive solution combination by applying a *dynamic updating* strategy within the SS *Reference Set Updating Method*. Moreover, since the algorithm fundament is on a combination of solutions, other relevant extension could be to consider the more and more employed strategy named as *Path-Relinking*. Likewise, the application of other hybrid approaches to the IR problem (*Ant Colony Optimization, other kinds of memetic algorithms*, etc.) are promising choices.

References

1. P. J. Besl and N. D. McKay: A method for registration of 3-D shapes. IEEE Transactions on Pattern Analysis and Machine Intelligence, 14: 239–256, 1992.
2. L. G. Brown: A survey of image registration techniques. ACM Computing Surveys, 24(4):325–376, 1992.
3. V. Campos, F. Glover, M. Laguna and R. Martí: An experimental evaluation of a scatter search for the linear ordering problem. Journal of Global Optimization, 21: 397–414, 2001.
4. O. Cordón, S. Damas, and J. Santamaría: A scatter search algorithm for the 3D image registration problem. In X. Yao et al., Eds., Parallel Problem Solving from Nature - PPSN VIII, 8th International Conference, Birmingham, UK, September 18-22. Lecture Notes in Computer Science 3242, Springer, 471–480, 2004.
5. O. Cordón, S. Damas, and J. Santamaría: A CHC evolutionary algorithm for 3D image registration. In T. Bilgic, B.D. Baets and O. Bogazici, Eds., Fuzzy Sets and Systems - IFSA 2003, 10th International Fuzzy Systems Association World Congress, Istanbul, Turkey, June 30 - July 2, 2003, Proceedings. Lecture Notes in Artificial Intelligence 2715, Heidelberg, Springer, 404–411, 2003.
6. O. Cordón, S. Damas, and E. Bardinet: 2D image registration with iterated local search. In J.M. Benítez, O. Cordón, F. Hoffmann, R. Roy, Eds., Advances in Soft Computing. Engineering Design and Manufacturing, Springer, 233–242, 2003.
7. J. Feldmar and N. Ayache: Rigid, affine and locally affine registration of free-form surfaces. International Journal of Computer Vision, 18(2):99–119, 1996.

8. S. Fernández-Vidal, E. Bardinet, S. Damas, G. Malandain, and N. Pérez de la Blanca: Object representation and comparison inferred from its medial axis. Proceedings of the International Conference on Pattern Recognition (ICPR'00), vol. 1, 712–715, Barcelona, Spain, 2000.

9. F. Glover: Heuristics for integer programming using surrogate constraints. Decision Sciences, 8: 156–166, 1977.

10. F. Glover: A template for scatter search and path relinking. Selected Papers from the Third European Conf. on Artificial Evolution, 3–54, Nimes, France, 1997.

11. F. Glover, M. Laguna, and R. Martí: Fundamentals of Scatter Search and Path Relinking. Control and Cybernetics, 39(3): 653–684, 2000.

12. D. E. Goldberg and R. Lingle Jr.: Alleles, loci, and the travelling salesman problem, Proc. of the First Intl. Conf. on Genetic Algorithms and Their Applications, Lawrence Erlbaum Associates Publishers, 1985.

13. K. Han, K. Song, E. Chung, S. Cho, and Y. Ha: Stereo matching using genetic algorithm with adaptive chromosomes. Pattern Recognition, 32:1729–1740, 2001.

14. R. He and P. A. Narayana: Global optimization of mutual information: application to three-dimensional retrospective registration of magnetic resonance images. Computerized Medical Imaging and Graphics, 26:277–292, 2002.

15. M. Laguna and R. Martí: Scatter Search: Methodology and Implementations in C. Kluwer Academic Publishers, Boston, 2003.

16. Y. Liu: Improving ICP with easy implementation for free-form surface matching. Pattern Recognition, 37: 211–226, 2004.

17. J. Luck, C. Little and W. Hoff: Registration of range data using a hybrid simulated annealing and iterative closest point algorithm. Proc. of IEEE Intl. Conf. on Rob. and Auto., 3739-3744, 2000.

18. O. Monga, S. Benayoun, and O. D. Faugeras: Using partial derivatives of 3D images to extract typical surface features. Proc. IEEE Computer Vision and Pattern Recognition (CVPR 92), Urbana Champaign, Illinois (USA), pp. 354–359, 1992.

19. P. Moscato: On Evolution, Search, Optimization, Genetic Algorithms and Martial Arts: Towards Memetic Algorithms. Technical Report, Caltech Concurrent Computation Program, C3P Report 826, 1989.

20. M.G.C. Resende and C.C. Ribeiro: Greedy randomized adaptive search procedures. In F. Glover and G. Kochenberger, Eds., State-of-the-art Handbook in Metaheuristics, Kluwer Academic Publishers, Boston, 219–250, 2001.

21. M.P. Wachowiak, R. Smolikova, Y. Zheng, J.M. Zurada, and A.S. Elmaghraby: An approach to multimodal biomedical image registration utilizing particle swarm optimization. IEEE Transactions on Evolutionary Computation 8(3): 289–301, 2004.

22. S. M. Yamany, M. N. Ahmed, and A. A. Farag: A new genetic-based technique for matching 3D curves and surfaces. Pattern Recognition, 32:1817–1820, 1999.

Evolution Strategies and Threshold Selection

Thomas Bartz-Beielstein

Department of Computer Science,
University of Dortmund, Germany
thomas.bartz-beielstein@udo.edu
http://ls11-www.cs.uni-dortmund.de/people/tom

Abstract. A hybrid approach that combines the $(1+1)$-ES and threshold selection methods is developed. The framework of the new experimentalism is used to perform a detailed statistical analysis of the effects that are caused by this hybridization. Experimental results on the sphere function indicate that hybridization worsens the performance of the evolution strategy, because evolution strategies are well-scaled hillclimbers: the additional threshold disturbs the self-adaptation process of the evolution strategy. Theory predicts that the hybrid approach might be advantageous in the presence of noise. This effect could be observed—however, a proper fine tuning of the algorithm's parameters appears to be advantageous.

1 Introduction

Following Stützle and Hoos, metaheuristic approaches can be described as generic techniques that are used "to guide or control an underlying problem-specific heuristic method in order to improve its performance or robustness"[1]. Hybrid metaheuristics combine methods of different metaheuristics. Two contradictory trends can be observed in recent research: (i) to develop more and more new algorithms or (ii) to analyze and understand existing heuristics and to add new features only when necessary. Following (ii), we will analyze potential assets and drawbacks that arise from a combination (hybridization) of evolution strategies and threshold selection. The analysis comprehends methods from the *new experimentalism*, that is an influential trend in recent philosophy of science. The new experimentalists develop statistical methods to set up experiments, to test algorithms, and to learn from the resulting errors and successes [2].

In many cases heuristics require the determination of parameters before the optimization run is performed. In the remainder of this paper, optimization runs will be treated as experiments. From the viewpoint of an experimenter, design variables (factors) are the parameters that can be changed during an experiment. Here comes the new experimentalism into play: a systematic variation of these factors and a statistical analysis of the resulting errors and successes are the keys for an understanding of the algorithm's performance. Generally, there are two different types of factors that influence the behavior of an optimization algorithm: (i) *problem specific* and (ii) *algorithm specific factors.*

M.J. Blesa et al. (Eds.): HM 2005, LNCS 3636, pp. 104–115, 2005.

These factors will be discussed in Section 2. Evolution strategies will be introduced in Section 3, and threshold selection approaches are presented in Section 4. Section 5 considers test problems and performance measures that are used afterwards to perform the experiments. The paper closes with a summary and conclusion.

2   Experimental Designs

Algorithm specific factors will be considered first: *Endogenous* can be distinguished from *exogenous* algorithm parameters. The former are kept constant during the optimization run, whereas the latter, e.g. standard deviations in evolution strategies, are modified by the algorithms during the run. An *algorithm design* $\mathcal{D}_A$ is a set of vectors with specific settings of an algorithm. A design can be specified by defining ranges of values for the design variables, e.g. "1:1:10" denotes integers from 1 to 10, whereas "1:10" denotes real numbers from the interval $[1, 10]$, or by specifying a set of values, e.g. "$\{1, 5, 10\}$". Note that a design can contain none, one, several or even infinitely many vectors. We will consider quantitative factors only. How qualitative factors can be included into the experimental analysis is discussed in [3].

Problem designs $\mathcal{D}_P$ provide information related to the optimization problem, such as the available resources, e.g. the number of function evaluations $t_{\max}$. Furthermore it is important to specify initialization and termination criteria. An *experimental design* $\mathcal{D}$ consists of a problem design $\mathcal{D}_P$ and an algorithm design $\mathcal{D}_A$. The run of a stochastic search algorithm can be treated as an experiment with a stochastic output $Y(x_a, x_p)$, with $x_a \in \mathcal{D}_A$ and $x_p \in \mathcal{D}_P$. If random seeds are specified, the output would be deterministic. This case will not be considered further, because it is not a common practice to specify the seed that is used in an optimization run.

Performance can be measured in many ways, for example as the best or the average function value from n runs (see also Section 5). One of our goals is to find a design point $x_a^* \in \mathcal{D}_A$ that improves the performance of an optimization algorithm for one problem design point $x_p \in \mathcal{D}_P$. To test the robustness of an algorithm, more than one design point can be considered. The approach to determine good design points presented in this paper is based on the *sequential parameter optimization* (SPO) methodology developed in [4] that has been applied successfully in several contexts, e.g. [5,6].

3   The Two Membered Evolution Strategy

The two membered evolution strategy, or $(1 + 1)$-ES, is included in our analysis for three reasons: (i) It is easy to implement, (ii) it requires only a few exogenous parameters, and (iii) it defines a standard for comparisons. Many optimization practitioners apply the $(1 + 1)$-ES (Figure 1) to their optimization problem. Schwefel [7] describes this algorithm as "the minimal concept for an imitation

---

Procedure $(1+1)$-ES/TS.

Initialization: Initialize the iteration counter: $t = 1$. Determine: (i) a point $X_1^{(t)}$
with associated position vector $x_1^{(t)} \in \mathbb{R}^d$, (ii) a standard deviation $\sigma^{(t)}$, and
(iii) a threshold value $\tau^{(t)}$. Determine the function value $y_1 = f(x_1^{(t)})$.
  while some stopping criterion is not fulfilled **do**
    repeat M times:
      Mutation: Generate a new point $X_2^{(t)}$ with associated position vector
      $x_2^{(t)}$ as follows:

$$x_2^{(t)} = x_1^{(t)} + z, \tag{1}$$

      where z is a d-dimensional vector. Each component of z is the real-
      ization of a normal random variable Z with mean zero and standard
      deviation $\sigma^{(t)}$.
      Evaluation: Determine the function value $y_2 = f(x_2^{(t)})$.
      Selection: Accept $X_2^{(t)}$ as $X_1^{(t+1)}$ if

$$y_2 + \tau^{(t)} < y_1, \tag{2}$$

      otherwise retain $X_1^{(t)}$ as $X_1^{(t+1)}$. Increment t.
    end.
    Adaptation:

$$\text{Update} \quad \sigma^{(t)}. \quad \text{Update} \quad \tau^{(t)}. \tag{3}$$

  done.

Fig. 1. The hybrid evolution/threshold selection strategy (ES/TS). The two membered
evolution strategy or $(1+1)$-ES for real-valued search spaces uses $M = 1$ and $\tau^{(t)} \equiv 0$.
The symbol f denotes an objective function $f : \mathbb{R}^d \to \mathbb{R}$ to be minimized. Threshold
selection (TS) uses a constant step-size $\sigma^{(t)} \equiv \sigma$ and a threshold adaptation scheme.

of organic evolution". The standard deviation σ will be referred to as *step-width*
or *mutation strength*. We will consider the following two ES-variants:

(ES-1) Constant Step Length. The basic $(1+1)$ algorithm does not modify
the step-size $\sigma^{(t)}$ in Equation 3 and uses a zero threshold $\tau^{(t)}$. It is expected
to be outperformed by other algorithms. However, sometimes unexpected
results may occur. Probably nothing unexpected may happen, "but if something did happen, that would be a stupendous discovery" [8]. This algorithm
requires the specification of a (constant) step-size $\sigma^{(t)}$ value only.

(ES-2) Step-Length Adaptation. Step-length adaptation relies on the following heuristic: *The step-size (standard deviation) should be adapted during the search. It should be increased, if many successes occur, otherwise
it should be reduced.* The 1/5 success rule derived by Rechenberg [9] while
analyzing the $(1+1)$-ES on two basically different objective functions for selecting appropriate step lengths can be seen as one instance of this heuristic:

From time to time during the optimization obtain the frequency of successes, i.e., the ratio of the number of the successes to the total number of trials (mutations). If the ratio is greater than 1/5, increase the variance, if it is less than 1/5, decrease the variance.

A more precise formulation is required to implement the 1/5 success rule. "From time to time during the optimization run" can be interpreted as "after every s_n mutations." "Increase the variation" can be implemented as a multiplication with a step-size adjustment factor. Other schemes are possible, e.g. to additive or exponential variations. The ratio of the number of the successes to the total number of mutations, the so-called *success rate* s_r, might be modified as well as the factor by which the variance is reduced or increased, the so-called *step size adjustment* factor s_a. We analyze the following two variants to implement the 1/5 rule:

(intv). A success counter $c \in \mathbb{N}_0$ is initialized at iteration $t = 1$. If a successful mutation occurs, c is increased. Every s_n iterations, the success rate is determined as c/s_n and c is set to zero.

(cont). A success vector $v^{(t)} \in \mathbb{B}^{s_n}$ is initialized at iteration $t = 1$: $v_k^{(t)} = 0$, $1 \leq k \leq s_n$. If a successful mutation occurs at iteration t, the $(1+t \mod s_n)$-th bit is set to 1, otherwise it is set to 0. After an initialization phase of s_n iterations, the success rate is determined in every iteration as $\sum_{k=1}^{s_n} v_k^{(t)}/s_n$.

The related algorithm designs are summarized in Table 1.

Table 1. Factors of the two membered evolution strategy. Based on the default values, the step size σ is multiplied by 0.85, if the success rate is larger than $1/s_r = 1/5$ or equivalently, if more than 20 out of 100 mutations have been successful.

Symbol	Factor	Range	Default
s_n	adaptation interval	$\mathbb{N}$	100
s_r	1/success rate	$\mathbb{R}_+$	5
s_a	step size adjustment factor	$\mathbb{R}_+$	0.85
$\sigma^{(0)}$	starting value of the step size σ	$\mathbb{R}_+$	1
$s_{1/5}$	step size update rule	{intv, cont }	cont

4   Threshold Selection Algorithms

Threshold rejection (TR) and *threshold acceptance* (TA) are complementary strategies. Threshold rejection has been proposed as a selection method for evolutionary algorithms, that accepts new candidates if their function values are significantly better than the values of the other candidates [10]. "Significant" is equivalent to "by at least a margin of τ". Threshold acceptance accepts a new candidate even if its function value is worse [11,12,13]. The term *threshold selection* (TS) subsumes both selection strategies. The hybrid approach presented in this paper analyzes how threshold selection can be integrated into the $(1+1)$-ES strategy (Figure 1). Threshold selection provides the opportunity to escape from local optima and is implemented in many algorithms, for example in simulated annealing: *During the initial iterations of a search algorithm it is advantageous to accept worse candidate solutions. The probability of accepting a worse candidate should be continuously reduced as the number of iterations increases.* However, it is a kind of art to choose a suitable annealing schedule [13]. The annealing schedule can be seen as one instance of a more general acceptance heuristic: *The probability of accepting a worse candidate solution should be adapted during the search. It should be reduced if a candidate solution is accepted, otherwise it should be increased.* We implemented three variants of threshold selection that have been integrated into evolution strategies. Table 2 summarizes the factors used in the threshold selection algorithms.

(TS-1) Constant. To integrate a threshold mechanism into an $(1 + 1)$-ES, a non-zero threshold value τ has to be determined. This threshold affects Equation 2.

(TS-2) Linear. This variant modifies the threshold value linearly. Negative threshold values values are increased during the search process as follows: $\tau^{(t)} = \tau(-1 + t/t_{\max})$, with $\tau^{(t)} \in [-\tau, 0]$. If positive threshold values are specified, the rule $\tau^{(t)} = \tau(1 - t/t_{\max})$, with $\tau^{(t)} \in [\tau, 0]$ is used to modify the threshold.

(TS-3) Self-adaptive. We integrated a self-adaptive annealing schedule into the algorithm. Although there is no obvious analogy for the "temperature" T with respect to the optimization problem, we will use T, because it is an established term to describe the variation of the acceptance probability during the search process. Let y_i denote the function values as defined in Figure 1. If the mutation was successful, the temperature T is modified according to $T = T/(1 + b\tau T)$, otherwise $T = T/(1 - \tau T)$. The new candidate solution is accepted with probability $\exp(\delta/T)$, with $\delta = y_2 - y_1$ (Equation 2). Note, that τ defines how much the temperature (that determines the probability of accepting a worse candidate solution) is decremented at each step as the cooling proceeds, and b specifies a balance factor.

Table 2. Factors of the threshold selection strategies. Note, that τ influences the acceptance probability in the self-adaptive threshold heuristic.

Symbol	Factor	Range	Default
τ	threshold value	$\mathbb{R}$	0
b	balance factor	$\mathbb{R}_+$	5
σ	value of the step size	$\mathbb{R}_+$	1

5   Experiments

Classical experimental approaches in evolutionary computation (i) define a set of test (standard) functions, (ii) run a certain number of algorithms, and (iii) finally compare the obtained results. The new experimentalism proposes a different methodology: (i) Formulate a set of questions (hypotheses or goals), (ii) select an appropriate set of test functions, (iii) run a certain number of algorithms, and (iv) search for environments in which these results cannot be repeated. However, the new experimentalism can benefit from the huge number of test functions available in the optimization literature. Besides standard measures to determine the algorithm's performance such as the average, median, minimum, maximum function values, and associated standard deviations, we report a measure based on bootstrap, that reflects the goals of optimization practitioners to select the best results from several runs and to skip the others:

1. Generate n results.
2. **repeat** k times:
   (a) Select (without replacement) a set M_i of $m < n$ values.
   (b) Determine $m_i := \min M_i$.
   end.
3. Calculate $\sum_i^k m_i/k$. The resulting value will be referred to as $\min_{\text{boot}}$.

The first goal of our experimental analysis is to find a suitable algorithm design x^*_{1+1} for the $(1+1)$ ES. The next goal is to find environments where this design does not work. The final analysis tries to find explanations, why special environments do not permit a generalization of the results found so far. If not stated otherwise, the methods used in this article do not require any assumption on the underlying distributions.

5.1   How to Determine a Good Algorithm Design?

Classical designs such as fractional factorial designs are used in this pre-experimental screening phase to eliminate worse algorithm designs. A very simple configuration, which uses the sphere function $\sum_i^d x_i^2$, was chosen first (Table 3). In the second step, more complex situations have been analyzed (varied starting points, increased dimension). Forthcoming papers will investigate more complex objective functions that introduce multi-modality or noise. Starting points have been initialized deterministically (DETEQ), the run terminated after $t_{\max}$ function evaluations (EXH), and the mean best function value from n runs was reported (MBST) [14]. The problem design $x^{(0)}_{\text{sphere}}$ from Table 3 was used to generate *run length distributions* (RLDs) [15]. The RLDs gave valuable hints to

Table 3. Problem design for the first pre-experimental experiments to determine a fair experimental setup: n denotes the number of repeated runs, $t_{\max}$ is the number of function evaluations, d the problem dimension, and $x^{(0)}$ is the starting point.

Design	n	$t_{\max}$	d	$x^{(0)}$
$x^{(0)}_{\text{sphere}}$	50	1000	1	100
$x^{(1)}_{\text{sphere}}$	50	1000	1	10:100
$x^{(2)}_{\text{sphere}}$	50	250	2	10:100
$x^{(3)}_{\text{sphere}}$	100	1000	$\{1, 2, 5, 10\}$	100
$x^{(4)}_{\text{sphere}}$	100	10^6	10:10:60	100

determine t_{max}, the maximum number of function evaluations for the comparisons and thereby to avoid floor and ceiling effects. These effects occur if the problem is too easy or too hard, respectively. The success limit was set to 10^{-6}, that means an optimization run was classified as successful, if it was able to determine a candidate solution x with $f(x) < 10^{-6}$.

The $(1+1)$-ES with algorithm design $x_{1+1}^{(0)}$ from Table 4 was chosen for this analysis. Designs from this table are used during the screening phase to detect outliers that can disturb the analysis. Note, interactions between factors can be more important than main factor effects [14]. An analysis of the RLDs from experiments that are based on algorithm design $x_{1+1}^{(2)}$ and problem design $x_{sphere}^{(0)}$ reveals that a budget of $t_{max} = 500$ function evaluations is sufficient. After 1000 function evaluations, only 50 percent of the runs with $s_a = 1$ attained the pre-specified function value (here: 10^{-6}), whereas 100 percent of the runs with $s_a = 0.9$ attained this border already after 300 function evaluations. This is a positive effect of the step-size adaptation on the performance: a step-size adjustment factor s_a of 1 keeps the step-width constant, whereas $s_a = 0.9$ enables an adaptation that is based on the success rate.

Table 4. $(1+1)$-ES: Algorithm designs to calibrate the experimental design and to avoid floor or ceiling effects. Threshold $\tau = 0$ and initial step size $\sigma^{(0)} = 1$ have been used.

Design	s_n	s_r	s_a
$x_{1+1}^{(0)}$	$\{10, 20, 100\}$	$\{2, 5, 10\}$	$\{0.5, 0.75, 0.9\}$
$x_{1+1}^{(1)}$	10	5	0.75
$x_{1+1}^{(2)}$	$\{10, 25\}$	5	$\{0.9, 1\}$
$x_{1+1}^{(3*)}$	2	7.25	0.758

SPO suggests to vary settings of problem design to guarantee that the observed effect was not caused by one specific situation. Therefore, we analyzed how algorithm designs scale with the problem dimension. Figure 2 depicts the relationship between problem dimension and the empirical cumulative density function of the number of function evaluations to reach a pre-specified goal. In

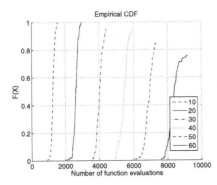

Fig. 2. Run-length distributions for 10-60 dimensional sphere functions. Increasing the problem dimension from d to $d+1$ requires approximately 100 additional function evaluations to obtain a similar solution. Algorithm design $x_{1+1}^{(1)}$ and problem design $x_{sphere}^{(4)}$.

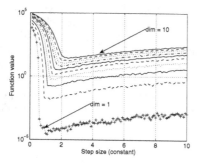 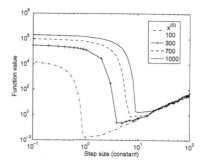

Fig. 3. Constant step length. Sphere function. Dimensions and starting points modified. Left: $x_{\text{sphere}}^{(6)}$ and x_{const}^*. Right: $x_{\text{sphere}}^{(7)}$ and x_{const}^*. Constant step-sizes of $\sigma = 1$ or $\sigma = 1.5$ appear to be useful.

a similar manner as the dimension was varied, different starting points have been used. The design $x_{\text{sphere}}^{(5)}$ with $t_{\max} = 500$, $d = 2$, and $x^{(0)} = 100$ has been determined in this pre-experimental phase and will be used for the following experiments.

5.2   A Comparison of Different Heuristics

The algorithms will be fine-tuned in this section to generate results that enable a fair comparison. In the first experiments, the algorithm with constant step-sizes (x_{const}) has been analyzed (Figure 3). Two variants of problem design $x_{\text{sphere}}^{(5)}$ have been used: $x_{\text{sphere}}^{(6)}$ varies problem dimensions (d=1:1:10), whereas $x_{\text{sphere}}^{(7)}$ uses different starting points $x^{(0)} \in \{100, 300, 700, 1000\}$.

The second series of experiments have been performed to analyze the influence of the success-rate determination scheme on the performance of the two-membered evolution strategy (problem design $x_{\text{sphere}}^{(5)}$ and the algorithm designs from Table 5). SPO will be used to fine-tune the algorithm design detected during the pre-experimental phase. A comparison of the RLDs shows only minor differences between the variants $x_{1+1}^{(4*)}$ and $x_{1+1}^{(5*)}$, e.g. $\min_{\text{boot}} = 5.30\text{e-}40$ and $1.28\text{e-}42$ respectively (Table 6). A plot of the observed difference [14] was used to analyze the statistical significance of their difference (Figure 4). We can conclude that there is a difference in means. If 50 (500) samples are drawn, this hypothesis would be wrong in 10 (1) out of 100 experiments. However: when A and B are different treatments with associated means μ_A and μ_B, μ_A and μ_B are certain to differ in some decimal places so that $\mu_A - \mu_B = 0$ is known in advance to be false. The observed difference is very small and large sample sizes (e.g. 500) are necessary for its statistical significance (i.e. to obtain a small p-value). Therefore we conclude that the observed difference is not scientifically meaningful. This is Step (S-12) as described in [4].

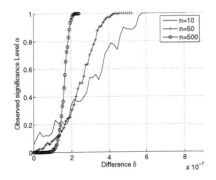

Fig. 4. Comparing the difference in the mean between success-rate schemes intv and cont. Designs $x_{\text{sphere}}^{(5)}$, $x_{1+1}^{(4*)}$, and $x_{1+1}^{(5*)}$ (Table 5). 500 samples are necessary to indicate that the hypothesis "there is a difference in means as large as $2 \cdot 10^{-7}$" would be wrong only in one out of 100 experiments. See [14] for a discussion of OSL-plots.

Why do the algorithms with the tuned designs $x_{1+1}^{(4*)}$ and $x_{1+1}^{(5*)}$ perform better than the default design? This means that algorithms with a very small memory, only two or seven bits, outperform algorithms with larger s_n values. Obviously, it takes s_n iterations to fill the memory vector. During this initial phase, no step-size adaption can occur. If the budget provides only $t_{\max} = 250$ iterations, a memory vector with more than 100 entries appears to be prohibitive. However, these considerations would explain small s_n values, but not extremely small values like $s_n = 2$. How the size of the memory vector affects the performance can be seen in Figure 5. It indicates that the $(1 + 1)$-ES is a well-scaled hill-climber. When big steps are advantageous, the algorithm takes big steps, and it takes little ones while approaching the optimum. The graph of the step size illustrates this behavior. A larger memory vector reacts too slowly, step-sizes should be adapted immediately.

Table 5. $(1 + 1)$-ES: Algorithm design to compare two success-rate determination schemes. Problem design $x_{\text{sphere}}^{(5)}$, $\tau = 0$ and $\sigma^{(0)} = 1$ for all experiments.

Design	s_n	s_r	s_a	$s_{1/5}$
$x_{1+1}^{(4)}$	$1:20$	$1:20$	$0.5:0.99$	cont
$x_{1+1}^{(5)}$	$1:20$	$1:20$	$0.5:0.99$	intv
$x_{1+1}^{(4*)}$	7	2.40	0.83	cont
$x_{1+1}^{(5*)}$	2	2.92	0.58	intv

Hybrid approaches have been considered next: Two factors, that are held constant during the optimization run, are necessary to specify the algorithm design of the first hybridization, that uses constant step-sizes and constant threshold values (x_{csct}): the step-size σ and the threshold τ. The experiments reveal that algorithms with zero threshold values perform best. Step sizes about 1 are preferred for the hybrid metaheuristic with constant step-sizes and constant threshold (x_{csct}). Results from the other hybridizations (constant step-sizes, linear threshold (x_{cslt}) and constant step-sizes with self-adaptive threshold (x_{sann})) indicate that non-zero threshold values worsen the algorithm's performance.

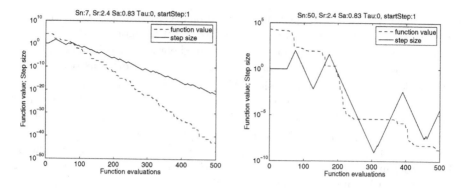

Fig. 5. The memory effect: the influence of different memory vector sizes on the search process. $s_r = 2.4$, $s_a = 0.831$, and $s_n = 7$ (left) or $s_n = 50$ (right). The plot of the logarithm of the function value over time in the left figure shows a straight line. The problem design $x^{(5)}_{\mathrm{sphere}}$ was used for these experiments.

Table 6. Experimental results from the hybridizations of ES and TS. Problem design $x^{(5)}_{\mathrm{sphere}}$ was used.

Algorithm	Mean	Median	Sd	Min	Max	min$_{\mathrm{boot}}$
x^{*}_{const}	1.65e-02	1.05e-02	1.78e-02	1.09e-05	1.46e-01	1.74e-05
$x^{(4*)}_{1+1}$	9.67e-27	2.07e-32	9.07e-26	4.98e-40	9.06e-25	5.30e-40
$x^{(5*)}_{1+1}$	2.73e-25	2.52e-34	2.57e-24	2.61e-44	2.57e-23	1.28e-42
x^{*}_{csct}	2.18e+03	399.43	4.89e+03	0.4656	1.89e+04	0.9612
x^{*}_{cslt}	0.3414	0.0164	2.6340	1.72e-06	25.9286	5.14e-04
x^{*}_{sann}	0.0160	0.0099	0.0171	5.38e-04	0.1013	5.82e-04
x^{*}_{τ}	0.0484	0.0410	0.0342	9.23e-04	0.1361	0.0012
$x^{*}_{\tau(t)}$	0.0474	0.0414	0.0340	8.54e-04	0.2327	0.0014
x^{*}_{temp}	7.21e-07	5.24e-07	7.95e-07	1.04e-08	5.67e-06	1.62e-08
NMS	1.15e-77	–	0	–	–	–

The next series of experiments have been set up to analyze whether constant or deterministically modified threshold schemes can improve the performance of evolution strategies. Experiments clearly indicate that non-zero threshold values worsen the algorithm's performance in this situation, too. Even the self-adaptive threshold scheme (x_{temp}) does not improve the algorithm's performance. The results from the experiments are summarized in Table 6. To improve comparability, results from a Nelder-Mead simplex (NMS) optimization have been added [16]. The Nelder-Mead algorithm requires the specification of four parameters (coefficients of reflection, expansion, contraction, and shrinkage), that have been tuned with SPO. The NMS optimization was able to find a candidate solution with function value 1.15e-77, a result that is significantly better then the ES/TS results. However, NMS fails completely on the sphere function in higher-dimensional cases, e.g. the best function value for $d = 50$ reads 4.77e+05, whereas the $(1+1)$-ES can cope with these problems (Fig. 2). This is an inherent problem of the NMS and not due to problems with the algorithm design.

Based on local performance measures, it can theoretically be shown that ES benefits from TS [10,14] under noise. Additive Gaussian noise has been added to the objective function from design $x_{\text{sphere}}^{(5)}$. The algorithm design $x_{1+1}^{(5*)}$, that has been tuned on the sphere function without noise, was used for the first experiments. Constant non-zero threshold values (TS-1) were able to improve the performance significantly. However, after applying SPO to $x_{1+1}^{(5*)}$ in the noisy environment, the tuned algorithm design performed better without threshold. These experiments indicate that there are situations (under noise), in which a combination of ES and TS might be beneficial.

6   Summary and Conclusion

The paper demonstrated the huge potential for the new experimentalism in computer science. Good algorithm designs can lead to impressive performance improvements and to robust algorithms that can be constructed systematically. SPO provides means for an in-depth understanding and fair comparisons of algorithms. The framework of the new experimentalism can be used to determine if statistically significant results are scientifically meaningful.

The SPO approach presented here can easily be applied to other algorithm–problem combinations. A recent paper discusses three scenarios to demonstrate its flexibility: (i) to analyze newly developed algorithms, (ii) to compare well-known algorithms, and (iii) to apply algorithms efficiently and effectively to complex real-world optimization problems [6]. Or, consider for example binary search spaces: mutation can be realized by random bit-flips of the position vector $x_1^{(t)}$. The probability p_m of flipping a bit can be regarded as the pendant to the mutation strength σ. Or, travelling salesperson problems can be regarded as ordering problems that require combinatorial search spaces. A search step operator defines the number of states n_s that can be reached from a parental state (neighborhood) within one move step. The number of move steps s can be seen as a pendant to the mutation strength σ.

The sphere function has been chosen as a test-function with a calculable influence on the results. Evolution strategies require only a small memory vector while optimizing the sphere—too much information (memory) is debilitating. No difference between the two step-size adaptation schemes (intv and cont) could be observed. Is this also true for higher dimensions and other test-functions?

ES clearly outperformed TS on the sphere due to its self-adaptiveness. Nevertheless, there may be other environments (problem designs), in which a hybrid approach is beneficial (noise, multi-modality, combinatorial optimization problems). Following the methodology presented in this paper, we are seeking for environments in which the step-size adaptation does not work and hybrid approaches can be improve the performance.

References

1. H. H. Hoos and T. Stützle, *Stochastic Local Search—Foundations and Applications.* Elsevier, 2005.
2. D. G. Mayo, *Error and the Growth of Experimental Knowledge.* The University of Chicago Press, 1996.
3. T. Bartz-Beielstein and S. Markon, "Tuning search algorithms for real-world applications: A regression tree based approach," in *Proc. 2004 Congress on Evolutionary Computation (CEC'04), Portland OR*, G. W. Greenwood, Ed., vol. 1. Piscataway NJ: IEEE Press, 2004, pp. 1111–1118.
4. T. Bartz-Beielstein, K. E. Parsopoulos, and M. N. Vrahatis, "Design and analysis of optimization algorithms using computational statistics," *Applied Numerical Analysis & Computational Mathematics (ANACM)*, vol. 1, no. 2, pp. 413–433, 2004.
5. C. Lasarczyk and W. Banzhaf, "Total synthesis of algorithmic chemistries," in *GECCO 2005: Proceedings of the Genetic and Evolutionary Computation Conference*, 2005, in print.
6. T. Bartz-Beielstein, C. Lasarczyk, and M. Preuss, "Sequential parameter optimization," in *Proc. 2005 Congress on Evolutionary Computation (CEC'05), Edinburgh.* Piscataway NJ: IEEE Press, 2005, in print.
7. H.-P. Schwefel, *Evolution and Optimum Seeking*, ser. Sixth-Generation Computer Technology. New York: Wiley Interscience, 1995.
8. I. Hacking, *Representing and intervening.* Cambridge University Press, 1983.
9. I. Rechenberg, *Evolutionsstrategie. Optimierung technischer Systeme nach Prinzipien der biologischen Evolution.* Stuttgart: frommann-holzboog, 1973.
10. S. Markon, D. V. Arnold, T. Bäck, T. Beielstein, and H.-G. Beyer, "Thresholding – A selection operator for noisy ES," in *Proc. 2001 Congress on Evolutionary Computation (CEC'01), Seoul*, J.-H. Kim, B.-T. Zhang, G. Fogel, and I. Kuscu, Eds. Piscataway NJ: IEEE Press, 2001, pp. 465–472.
11. J. Matyáš, "Random Optimization," *Automation and Remote Control*, vol. 26, no. 2, pp. 244–251, 1965.
12. E. C. Stewart, W. P. Kavanaugh, and D. H. Brocker, "Study of a global search algorithm for optimal control," in *Proceedings of the 5th International Analogue Computation Meeting, Lausanne*, Aug.-Sept. 1967, pp. 207–230.
13. G. Dueck and T. Scheuer, "Threshold accepting: a general purpose optimization algorithm appearing superior to simulated annealing," *Journal of Computational Physics*, vol. 90, pp. 161–175, 1990.
14. T. Bartz-Beielstein, "New experimentalism applied to evolutionary computation," Ph.D. dissertation, University of Dortmund, April 2005.
15. H. H. Hoos, "Stochastic local search – methods, models, applications," Ph.D. dissertation, Technische Universität Darmstadt, 1998.
16. J. C. Lagarias, J. A. Reeds, M. H. Wright, and P. E. Wright, "Convergence properties of the Nelder–Mead simplex method in low dimensions," *SIAM J. on Optimization*, vol. 9, no. 1, pp. 112–147, 1998.

A Hybrid GRASP with Data Mining for the Maximum Diversity Problem

L.F. Santos, M.H. Ribeiro, A. Plastino*, and S.L. Martins**

Universidade Federal Fluminense, Departamento de Ciência da Computação
Niterói, RJ, Brazil
{lsantos, mribeiro, plastino, simone}@ic.uff.br

Abstract. The maximum diversity problem (MDP) consists in identifying, in a population, a subset of elements, characterized by a set of attributes, that present the most diverse characteristics among themselves. The identification of such solution is an NP-hard problem. In this work, we propose a hybrid GRASP metaheuristic for the MDP that incorporates a data mining process. Data mining refers to the extraction of new and potentially useful knowledge from datasets in terms of patterns and rules. We believe that data mining techniques can be used to extract patterns that represent characteristics of sub-optimal solutions of a combinatorial optimization problem. Therefore these patterns can be used to guide the search for better solutions in metaheuristics procedures. Performance comparison between related work and the proposed hybrid heuristics is provided. Experimental results show that the new hybrid GRASP is quite robust and, mainly, this strategy is able to find high-quality solutions in less computational time.

1 Introduction

Given a set of n elements $P = \{1, \ldots, n\}$ and a number m, $1 < m < n$, the maximum diversity problem (MDP) [4,5,6] consists in identifying a subset M with m elements from the population P, so that these elements present the maximum possible diversity among them.

Let i_k, $k \in L = \{1, \ldots, l\}$, be the l attributes associated with each element i. The measure of diversity d_{ij} between a pair of elements (i, j) is calculated by some function applied on their attributes. The choice of this function depends on the context of the problem instantiation. The MDP can then be formulated as: Maximize $z = \sum_{i=1}^{n-1} \sum_{j=i+1}^{n} d_{ij} x_i x_j$, subject to $\sum_{i=1}^{n} x_i = m$, where x_i is a binary variable indicating if an element i is selected to be a member of M.

Many applications [10] can be solved using the resolution of this problem, such as human resource management, measure of biodiversity, and VLSI design.

This problem belongs to the class of NP-hard problems as shown by Glover et al. [5]. In this work they also presented a mixed integer zero-one formulation for this problem, used to solve small instances by exact methods.

---

* Work sponsored by CNPq research grants 300879/00-8 and 475124/03-0.
** Work sponsored by CNPq research grant 475124/03-0.

M.J. Blesa et al. (Eds.): HM 2005, LNCS 3636, pp. 116–127, 2005.

Since the MDP is an NP-hard problem, some heuristics were developed to obtain approximate solutions. Constructive and destructive heuristics were presented by Glover et al. [6], which were evaluated using instances with different population sizes (maximum value was 30). The proposed heuristics obtained results close (2%) to the ones obtained by the exact algorithm, but much faster. Weitz and Lakshminarayanan [17] developed heuristics to find groups of students with the most possible diverse characteristics, such as nationality, age, and graduation level. They tested the heuristics using instances based on real data and implemented an exact algorithm for solving them.

Many works have been developed using the metaheuristic GRASP (Greedy Randomized Adaptive Search Procedure) [3] to solve the MDP. At first, good results for small instances of the problem were obtained by Ghosh [4]. Andrade et al. [1] developed a new construction phase for this GRASP and showed results for instances randomly created with a maximum population of 250 individuals. This proposal was able to find solutions better than the ones found by Ghosh's algorithm. In [2], Andrade et al. showed that incorporating path-relinking to the pure GRASP developed in [1] enabled to achieve better results. Silva et al. [15] elaborated new construction and local search heuristics and combined them to generate several GRASP algorithms. In order to evaluate these strategies, instances were created by the authors for populations of maximum size equal to 500 individuals. They compared their algorithms with the ones developed by Ghosh [4] and Andrade et al. [1] and showed that better results were achieved.

In this work, we propose a hybrid GRASP heuristic for the MDP, which incorporates a data mining process. We believe that data mining techniques can be used to extract patterns that represent characteristics of sub-optimal solutions of a combinatorial optimization problem. Therefore these patterns can be used to guide the search for better solutions in metaheuristics procedures. In [14], we obtained good results with this kind of hybridization for the Set Packing Problem. Basically, after GRASP executes a significant number of iterations, the data mining process extracts patterns from an elite set of solutions that guide the following GRASP iterations. Solutions are represented by sets of items and patterns are defined as subsets of items that occurs in a significant number of solutions. The process of mining these patterns is the well known problem called frequent itemset mining, a sub-problem of association rule mining [7,9]. The GRASP used to develop the hybrid strategy is based on the KLD strategy, a state-of-the-art algorithm for the MDP presented by Silva et al. [15].

In Sec. 2, we describe the KLD GRASP heuristic for the MDP presented in [15]. Data mining concepts are presented in Sec. 3. The hybridization of GRASP, proposed in this work, is defined in Sec. 4. The experimental results are reported and discussed in Sec. 5. Concluding remarks are made in Sec. 6.

2   GRASP Heuristic for the MDP

GRASP [3] is an iterative process, where each iteration consists of two phases: construction and local search. A feasible solution is built in the construction phase, and then its neighborhood is explored by the local search. The result is the best solution found over all iterations.

The construction phase of GRASP is an iterative process where, at each iteration, the elements $c \in C$, which are the elements that do not belong to the solution, are evaluated by a greedy function $g : C \rightarrow \Re_+$, that estimates the gain of including it in the partial solution. They are ordered by their estimated value in a list called restricted candidate list (RCL) and one of them is randomly chosen and included in the solution. The size of the RCL is limited by a parameter α. For a maximization problem, only the elements whose g values are in the range $[(1 - \alpha)g_{max}, g_{max}]$ are placed in RCL, where g_{max} is the maximum value of g obtained for all elements in C. This iterative process stops when a feasible solution is obtained.

The solutions generated by the construction phase are not guaranteed to be locally optimal. So, a local search is performed to attempt to improve each constructed solution. It works by successively replacing the current solution by a better one from its neighborhood, until no more better solutions are found.

Next, we describe the pure GRASP proposed by Silva et al. [15], called KLD, which is the algorithm selected to be hybridized with data mining techniques.

2.1   KLD Algorithm

Usually, the local search phase demands great computational effort and execution time, so the construction phase plays an important role to diminish this effort by supplying good starting solutions for the local search. The KLD algorithm is defined by a construction procedure developed by Silva et al. [15] which is combined with the local search developed by Ghosh [4].

The construction procedure is based on two existing techniques: filtering of constructed solutions [13] and reactive GRASP proposed in [12].

Filtering of constructed solutions leads to a more greedy construction. For each GRASP iteration, the construction algorithm is executed several times generating different solutions and only the best solution is selected to be used as the initial solution for the local search phase.

Prais and Ribeiro [12] proposed the procedure called reactive GRASP, for which the parameter α used in the construction phase is self adjusted for each iteration. The aim is to associate higher probabilities to values of α that lead to better solutions and lower ones to values of α that guide to worse solutions.

The KLD (K Largest Distances) algorithm uses both techniques described before. This algorithm constructs an initial solution by randomly selecting an element from a RCL of size K at each construction iteration. The RCL is created by first computing, for each element $i \in P$, the sum s_i of the diversity values $d_{ij}, j \in P \backslash \{i\}$, then selecting the K elements i that exhibit the largest values of s_i.

The procedure developed to implement the reactive GRASP starts considering m_it to be the total number of GRASP iterations. In the first block of iterations $B_1 = 0.4m\_it$, four different values for $K \in \{K_1, K_2, K_3, K_4\}$ are evaluated by dividing the block into four equal intervals $c_i, i = 1, \ldots, 4$. The value K_i is used for all iterations belonging to interval c_i. The values of K_i are shown in Table 1, where $\mu = (n - m)/2$. After the execution of the last iteration of block B_1, the quality of the solutions obtained for each K_i is evaluated. The average diversity value $zm_i = \sum_{1 \leq q \leq 0.1m\_it} z(sol_{iq})$ for the solutions $sol_{iq}, i = 1, \ldots, 4; q = 1, \ldots, 0.1m\_it;$ is obtained for each K_i. The values K_i are stored in a list $LK = \{lk_i\}, i = 1, \ldots, 4$, ordered by their zm_i values.

Table 1. K values for block B_1

i	c_i	K
1	$[1, \ldots, 0.1m\_it]$	$m + \mu - 0.2\mu$
2	$(0.1m\_it, \ldots, 0.2m\_it]$	$m + \mu - 0.1\mu$
3	$(0.2m\_it, \ldots, 0.3m\_it]$	$m + \mu + 0.1\mu$
4	$(0.3m\_it, \ldots, 0.4m\_it]$	$m + \mu + 0.2\mu$

Then the next block of iterations $B_2 = 0.6m\_it$ is divided into four intervals y_i, each one with different number of iterations. For each interval a value for K is adopted as shown in Table 2. In this way, the values K_i that provide better solutions are used in a larger number of iterations.

Table 2. K values for block B_2

i	y_i	K
1	$(0.4m\_it, \ldots, 0.64m\_it]$	lk_1
2	$(0.64m\_it, \ldots, 0.82m\_it]$	lk_2
3	$(0.82m\_it, \ldots, 0.94m\_it]$	lk_3
4	$(0.94m\_it, \ldots, m\_it]$	lk_4

At each GRASP iteration, the filter technique is applied by constructing two solutions and only the best solution is sent to the local search procedure.

The pseudo-code, including the description of the procedure for the construction phase using K largest distances heuristic, is given in Fig. 1.

In line 1, the cost of the best solution found in the execution of max_sol_filter iterations is initialized. The value K to be used to build the restricted candidate list (RCL) is calculated by the procedure det_K in line 2. This procedure defines the value for K, implementing the reactive GRASP described before. In line 3, the RCL is built. From line 4 to line 15, the construction procedure is executed max_sol_filter times and only the best solution is returned to be used as an initial solution by the local search procedure. From line 6 to line 10, a solution is constructed by the random selection of an element from RCL. In lines 11 to 14, the best solution found by the construction procedure is updated. If the GRASP iteration belongs to B_1 block, the cost of the solution found using the selected

```
procedure constr_KLD(it_GRASP, m_it, max_sol_filter, n, m)
 1.    best_cost_sol ← 0;
 2.    K ← det_K(it_GRASP, m_it, LK);
 3.    RCL ← Build_RCL(K, n);
 4.    for j = 1, ..., max_sol_filter do
 5.      sol ← {};
 6.      for k = 1, ..., m do
 7.        Randomly select an individual e* from RCL;
 8.        sol ← sol ∪ {e*};
 9.        RCL ← RCL − {e*};
10.      end for;
11.      if (z(sol) > best_cost_sol) then do
12.        sol_constr ← sol;
13.        best_cost_sol ← z(sol);
14.      end if
15.    end for;
16.    if (it_GRASP < 0.4m_it) then do
17.      Update_Sol_K(K, sol_eval, z(sol_constr));
18.    end if;
19.    if (it_GRASP == 0.4m_it) then do
20.      LK ← Build_LK(sol_eval);
21.    end if;
22.    return sol_constr.
```

Fig. 1. Construction procedure used to implement the KLD heuristic

K is stored in line 17. When the first block B_1 of iterations ends, the values K_i are evaluated and put in the list LK sorted in descending order, in line 20.

The local search strategy employed is the same as the one developed by Ghosh [4] and is described next. The neighborhood of a solution is the set of all solutions obtained by replacing an element in the solution by another that does not belong to the set of its elements. The incumbent solution M is initialized with the solution obtained by the construction phase. For each $i \in M$ and $j \in P \setminus M$, the improvement due to exchanging i by j, $\Delta z(i, j) = \sum_{u \in M_{\{i\}}} (d_{ju} - d_{iu})$ is computed. If for all i and j, $\Delta z(i, j) < 0$, the local search is terminated, because no exchange will improve z. Otherwise, after the elements of the pair (i, j) that provides the maximum $\Delta z(i, j)$ are interchanged, a new incumbent solution M is created and the local search is performed again.

3   Data Mining

Data mining refers to the extraction of new and potentially useful knowledge from datasets in terms of patterns and rules. Data mining processes have been used in a large variety of fields such as marketing, finance, health care, education, and security. The main kinds of rules and patterns mined from datasets are: association rules, sequential patterns, classification rules, and data clusters [9].

Association rules describe relationships among data items of a specific knowledge domain, which are hidden in large datasets. Market basket analysis is a typical application of this technique and consists in identifying relationships among products that significantly occur in customers purchases.

An association rule, defined over a set of items $\mathcal{I} = \{i_1, i_2, \ldots, i_n\}$, is an implication of the form $X \Rightarrow Y$, where $X \subset \mathcal{I}$, $Y \subset \mathcal{I}$, $X \neq \emptyset$, $Y \neq \emptyset$, and $X \cap Y = \emptyset$. Let $\mathcal{D}$ be a set of transactions defined over $\mathcal{I}$, where each transaction t is a subset of $\mathcal{I}$ ($t \subseteq \mathcal{I}$). Then, the rule $X \Rightarrow Y$ holds in database $\mathcal{D}$ with support s and confidence c if, respectively, $s\%$ of the transactions in $\mathcal{D}$ contain $X \cup Y$, and $c\%$ of the transactions in $\mathcal{D}$ that contain X also contain Y.

The problem of extracting association rules from a transactional database is commonly broken into two phases. Let $minsup$ and $minconf$ be, respectively, the user specified minimum support and confidence. The first phase, which characterizes the frequent itemset mining (FIM) problem, consists in identifying all frequent itemsets, i.e., sets of items that occur in at least $minsup\%$ of the database transactions. A frequent itemset is called maximal if it has no superset that is also frequent. The second phase outputs, for each identified frequent itemset Z, all association rules $A \Rightarrow B$ with confidence greater or equal to $minconf$, such that $A \subset Z$, $B \subset Z$, and $A \cup B = Z$. The FIM phase demands more computational effort than the second one and has been intensively addressed [7].

In this work, the useful patterns to be mined are sets of elements that commonly appear in sub-optimal solutions of the MDP. In this frequent itemset mining application, the set of items $\mathcal{I} = \{i_1, i_2, \ldots, i_n\}$ is the population P of the MDP. Each transaction t of the database $\mathcal{D}$ represents a sub-optimal solution of the MDP. A frequent itemset mined from $\mathcal{D}$ with support $s\%$ represents a set of elements that occur in $s\%$ of the elite solutions.

In order to mine the maximal frequent itemsets to be used in the proposed hybrid GRASP, we have used the FPmax* algorithm[1] [8].

4   The Hybrid GRASP for the MDP

In this section, we describe the strategy used to hybridize the GRASP metaheuristic to incorporate a data mining process. The basic procedure is composed by two phases. The first one (elite set generation phase) consists in executing pure GRASP iterations for n_iter iterations to obtain a set $\mathcal{S}$ of different solutions. Let $\mathcal{D}$ be the set containing the d best solutions of $\mathcal{S}$. To extract a set of patterns R from this elite set of solutions we apply an algorithm to mine the frequent itemsets from $\mathcal{D}$. Next, a second phase of GRASP (hybrid phase) is executed also for n_iter iterations using an adapted construction phase that builds a solution based on some pattern $p \in R$, i.e., each constructed solution will contain the elements of a pattern $p \in R$. The local search for each iteration is the same as that described in Subsec. 2.1.

The pseudo-code of the new adapted construction procedure is presented in Fig. 2. The parameter p will guide the construction of the initial solution which

---

[1] Available at *http://fimi.cs.helsinki.fi*.

```
procedure new_constr_KLD(it_GRASP, m_it, max_sol_filter, n, m, p)
  1.   best_cost_sol ← 0;
  2.   K ← det_K(it_GRASP, m_it, LK);
  3.   RCL ← Build_RCL(K, n);
  4.   for j = 1, ..., max_sol_filter do
  5.     sol ← p;
  6.     for k = 1, ..., (m − |p|) do
  7.       Randomly select an individual e* from RCL;
  8.       sol ← sol ∪ {e*};
  9.       RCL ← RCL − {e*};
 10.     end for;
 11.     if (z(sol) > best_cost_sol) then do
 12.       sol_constr ← sol;
 13.       best_cost_sol ← z(sol);
 14.     end if
 15.   end for;
 16.   if (it_GRASP < 0.4m_it) then do
 17.     Update_Sol_K(K, sol_eval, z(sol_constr));
 18.   end if;
 19.   if (it_GRASP == 0.4m_it) then do
 20.     LK ← Build_LK(sol_eval);
 21.   end if;
 22.   return sol_constr.
```

Fig. 2. New adapted construction procedure for GRASP heuristic for MDP

will contain all elements of p. In lines 2 and 3, the value of K and the RCL are defined as in the original construction algorithm, except that the RCL does not contain the elements in p. As in the algorithm showed in Subsec. 2.1, the construction phase is executed max_sol_filter times, and only the best solution found is sent to the local search. In line 5, the initial solution is defined with all elements of p. From lines 6 to 10 we apply the same procedure presented in lines 6 to 10 of Fig. 1 to obtain the other $m − |p|$ elements of the solution. At the end, the constructed solution contains all elements of p and some other elements selected by the construction procedure. In lines 11 to 21, the same procedures presented in lines 11 to 21 in Fig. 1 are performed to update the better solution and to set the values of K.

The data mining strategy adopted to extract a set of patterns R from the elite set of solutions $\mathcal{D}$ uses the algorithm FPmax* to mine the maximal frequent itemsets from $\mathcal{D}$. This algorithm is executed $d−1$ times, each one using a different value for support $s \in S = \{2, ..., d\}$, where $d = |\mathcal{D}|$. All patterns generated by each execution are merged in a set F and the r largest size patterns obtained from F are selected to generate R. The iterations of the GRASP second phase are divided into r intervals, and, for each interval, one different pattern $p \in R$ is used for the construction procedure. As defined in Sec. 3, a frequent itemset is called maximal if it has no superset that is frequent. Adopting maximal frequent itemsets we guarantee that a pattern is not a subset of another one leading to a more effective diversification.

According to the taxonomy of hybrid metaheuristics proposed in [16], our work could be classified as a high-level and relay hybrid metaheuristic. It is considered high-level since the data mining technique and the GRASP are self-contained and it is a relay hybridization because GRASP, the data mining techniques, and GRASP again are applied in a pipeline fashion.

5   Computational Results

The computational experiments were performed on a set of different test problems with distinct characteristics [15]. We selected from this work the larger and more difficult instances to solve. Each one consists of a population P with size n and a diversity matrix, which contains the diversity d_{ij} between elements i and j of P. The instances have populations sizes $n = 200, 300, 400, 500$. We have also produced four new instances with population size $n = 600$, using the generator developed in [15]. For all instances, tests were performed to find subsets M of sizes 10%, 20%, 30%, and 40% of the population size.

We obtained the source code of KLD from its authors, which is an improved version of KLD presented in [15]. The hybrid version was implemented incorporating the data mining technique into this same code.

The algorithms were implemented in C and compiled with `gcc 2.96`. The tests were performed on a 1.7 GHz Intel Pentium 4 with 256 Mbytes of RAM.

In Table 3, we show the CPU times observed for the KLD GRASP and the hybrid GRASP (DM-GRASP) using all instances. Both strategies executed 500 iterations. The first and second phases of DM-GRASP executed 250 iterations each one. The elite set $\mathcal{D}$, from which patterns were mined, had size 10, and the set of patterns R had size 10.

The first and second columns identify the parameters of each instance: the size n of the population and the number m of elements to be selected. For all instances, each procedure was executed ten times with different random seeds. The third and sixth columns show the average time (in seconds) for the KLD and DM-GRASP strategies, respectively. For these average values, the bold value indicates the best behavior, represented by the smaller amount of time. The fourth and seventh columns present the time (in seconds) of the fastest execution of KLD and DM-GRASP, among the ten tests. The fifth and eighth columns show the slowest executions of each strategy.

Based on the average values, we observe that DM-GRASP executed faster than KLD heuristic in all tests. The relative difference ranged from 2.3% to 10.0%. On average, DM-GRASP was 6.8% faster than KLD. Indeed, in DM-GRASP, the adapted construction procedure of the second phase takes less time to execute because some elements are already set in the solution (elements from the mined patterns). Then a smaller number of elements must be processed and inserted into the constructed solution. Besides, we suppose that the adapted construction procedure of DM-GRASP generates better initial solutions so that the local search takes less time to converge to a local optimal solution. We also observe in Table 3 that, in 18 instances, the slowest execution time of DM-GRASP (eighth column) is lower than the average time of the KLD (third column).

Table 3. Average, best and worst computational times for KLD and DM-GRASP

Instance		KLD			DM-GRASP		
n	m	Avg.	Best	Worst	Avg.	Best	Worst
200	20	32.9	30.1	39.2	**30.8**	30.3	30.9
200	40	136.0	133.0	142.7	**130.5**	125.8	145.7
200	60	345.7	340.5	371.4	**332.0**	323.3	347.1
200	80	555.0	545.0	563.7	**528.7**	519.6	537.9
300	30	223.0	210.5	237.9	**212.1**	203.9	220.2
300	60	928.9	903.3	956.6	**859.4**	844.0	883.4
300	90	2041.1	1992.5	2133.5	**1867.1**	1842.0	1907.7
300	120	3273.2	3214.7	3344.2	**3062.0**	2960.0	3119.8
400	40	807.2	795.5	819.4	**775.8**	760.9	782.0
400	80	3403.7	3340.1	3441.4	**3090.1**	3035.7	3221.5
400	120	7784.3	7669.1	7996.2	**7006.4**	6888.2	7160.4
400	160	11170.1	10928.4	11440.3	**10066.2**	9867.7	10314.9
500	50	1997.9	1968.3	2036.9	**1940.8**	1926.4	1965.3
500	100	9206.0	8986.4	9302.2	**8343.5**	8203.2	8463.6
500	150	19688.6	19500.6	19911.8	**17742.1**	17481.7	18062.9
500	200	29127.0	28576.3	29422.4	**26366.0**	25393.0	26971.0
600	60	4184.4	4151.3	4225.3	**4089.1**	4044.5	4132.7
600	120	17710.9	17569.0	17895.5	**16236.8**	16073.5	16543.5
600	180	40139.3	39802.1	40525.5	**37109.8**	36053.4	37989.4
600	240	59631.4	59370.9	59997.5	**55743.2**	55346.6	56888.7

In Table 4, we show the quality results for both KLD and DM-GRASP obtained over the same tests. The first and second columns identify the parameters of each instance. For all tests, each procedure was executed ten times. The third and sixth columns show the average diversity values obtained by KLD and DM-GRASP strategies. For these average values, when two different results are reached, a bold value indicates the best average found. The fourth and seventh columns present the standard deviation of the obtained diversity values. The fifth and eighth columns show the best diversity values achieved by both KLD and DM-GRASP strategies. In order to compare the obtained results to those reached by the other proposed strategies in the literature, the last column shows the best known value for each test. When different best diversity values are reached, an underlined value indicates the best one. The instances with population size 600 are new and the best known values are not indicated.

Considering the average values of the 20 tests, the two algorithms obtained the same result in 4 cases. In those 16 tests in which different average diversity values were reached, DM-GRASP heuristic found better results than KLD in 10 tests. The standard deviation values related to DM-GRASP executions were less or equal to the ones of KLD in 14 instances. In 11 cases, from the 16 tests where a best known result is shown, both algorithms were able to find the best known value. For one instance they did not reach the best result of the literature and, in three instances, they found new best values. In one case, only the DM-GRASP strategy reached the best known value. These results show that the new

Table 4. Average and best diversity values obtained by KLD and DM-GRASP

Instance		KLD			DM-GRASP			Best known
n	m	Avg.	S.D.	Best	Avg.	S.D.	Best	
200	20	1246.6	1.3	1247	1246.6	0.6	1247	1247
200	40	**4450.0**	0.0	4450	4449.0	0.9	4450	4450
200	60	9437.0	0.0	9437	9437.0	0.0	9437	9437
200	80	16225.0	0.0	16225	16225.0	0.0	16225	16225
300	30	**2688.4**	4.1	2694	2686.8	2.3	2694	2694
300	60	9680.9	8.4	9689	**9685.6**	2.8	9689	9689
300	90	**20735.9**	6.0	20743	20735.2	6.0	20743	20734
300	120	35880.3	1.2	35881	**35880.4**	1.0	35881	35881
400	40	**4652.0**	2.6	4658	4651.6	4.7	4658	4658
400	80	16941.5	12.7	16956	**16943.4**	10.6	16956	16956
400	120	36300.6	12.1	36317	**36304.0**	7.8	36317	36315
400	160	62455.3	15.0	62475	**62467.8**	11.2	62483	62483
500	50	**7117.8**	6.2	7130	7116.9	8.3	7129	7131
500	100	26248.5	7.9	26258	**26251.8**	8.2	26258	26254
500	150	56571.9	0.3	56572	56571.9	0.3	56572	56572
500	200	97330.0	10.1	97344	**97332.5**	10.0	97344	97344
600	60	10120.1	15.4	10150	**10123.5**	13.3	10149	new instance
600	120	37007.7	15.6	37035	**37024.8**	21.6	37058	new instance
600	180	80392.5	12.8	80410	**80394.3**	10.0	80407	new instance
600	240	**139028.5**	11.8	139048	139024.0	13.7	139040	new instance

hybrid GRASP is quite robust. The DM-GRASP is competitive when compared to KLD, a state-of-the-art algorithm for the MDP. Besides, this strategy is able to find high-quality solutions in less computational time.

Since DM-GRASP was always faster than KLD, to conduct a fair comparison between them, we let DM-GRASP run the same time KLD took for each different combination of instance and random seed. Table 5 shows that DM-GRASP outperforms KLD when both are allowed to run the same execution time. Based on the average values: they reached the same result in 2 instances, DM-GRASP found better results in 13 tests, and KLD performed better than DM-GRASP in just 5 instances.

6   Concluding Remarks

In this work, we proposed a hybrid GRASP metaheuristic that incorporates a data mining process for solving the maximum diversity problem.

One of our goals was to evidence that using a data mining technique can improve results obtained by GRASP. Computational experiments showed that employing frequent patterns mined from an elite set in the construction phase conducted GRASP to achieve better solutions for the MDP. The time analysis showed that the data mining strategy accelerates the process of finding a good

Table 5. Average and best diversity values obtained by KLD and DM-GRASP executing the same time

Instance		KLD			DM-GRASP			Best known
n	m	Avg.	S.D.	Best	Avg.	S.D.	Best	
200	20	**1246.6**	1.3	1247	1246.0	2.2	1247	1247
200	40	**4450.0**	0.0	4450	4449.4	1.0	4450	4450
200	60	9437.0	0.0	9437	9437.0	0.0	9437	9437
200	80	16225.0	0.0	16225	16225.0	0.0	16225	16225
300	30	**2688.4**	4.1	2694	2687.9	5.7	2694	2694
300	60	9680.9	8.4	9689	**9685.7**	2.7	9689	9689
300	90	**20735.9**	6.0	20743	20735.2	6.0	20743	20734
300	120	35880.3	1.2	35881	**35880.4**	1.0	35881	35881
400	40	4652.0	2.6	4658	**4652.2**	3.2	4658	4658
400	80	16941.5	12.7	16956	**16943.8**	10.4	16956	16956
400	120	36300.6	12.1	36317	**36306.3**	8.1	36317	36315
400	160	62455.3	15.0	62475	**62467.8**	11.2	62483	62483
500	50	7117.8	6.2	7130	**7119.5**	7.8	7130	7131
500	100	26248.5	7.9	26258	**26252.4**	7.1	26258	26254
500	150	56571.9	0.3	56572	**56572.0**	0.0	56572	56572
500	200	97330.0	10.1	97344	**97333.1**	9.8	97344	97344
600	60	10120.1	15.4	10150	**10123.5**	13.4	10149	new instance
600	120	37007.7	15.6	37035	**37025.9**	20.1	37058	new instance
600	180	80392.5	12.8	80410	**80399.9**	9.2	80411	new instance
600	240	**139028.5**	11.8	139048	139025.1	13.1	139040	new instance

solution by the GRASP procedure. Under the same execution time limit, the hybrid GRASP performed even better than pure GRASP.

In the hybrid proposal, the used patterns are mined just once. We believe that an improvement may come from mining the patterns not only once during the whole process, but alternating GRASP iterations and the mining procedure, which may allow the extraction of patterns from refined elite solutions.

We believe that the pattern size may influence the results, so we intend to improve the hybrid strategy defining a policy to choose an adequate pattern size.

Another future work can be a cooperative parallelization of GRASP. This approach would allocate a processor dedicated to mine patterns from the solutions generated by the other processors executing hybrid GRASP iterations.

References

1. Andrade, P. M. F., Plastino, A., Ochi, L. S., Martins, S. L.: GRASP for the maximum diversity problem, Procs. of MIC 2003, CD-ROM Paper: MIC03_15, (2003).
2. Andrade, M. R. Q., Andrade, P. M. F., Martins, S. L., Plastino, A.: GRASP with path-relinking for the maximum diversity problem, 4th Intl. Workshop on Experimental and Efficient Algorithms, LNCS **3503** (2005), 558–569.
3. Feo, T. A., Resende, M. G. C.: Greedy randomized adaptive search procedures, Journal of Global Optimization **6** (1995), 109–133.

4. Ghosh, J. B.: Computational aspects of the maximum diversity problem, Operations Research Letters **19** (1996), 175–181.
5. Glover, F., Hersh, G., McMillan, C.: Selecting subsets of maximum diversity, MS/IS Report No. 77-9, University of Colorado at Boulder (1977).
6. Glover, F., Kuo, C-C., Dhir, K. S.: Integer programming and heuristic approaches to the minimum diversity problem, J. of Bus. and Management **4** (1996), 93–111.
7. Goethals, B., Zaki, M. J.: Advances in frequent itemset mining implementations: Introduction to FIMI'03. IEEE ICDM FIMI Workshop (2003).
8. Grahne, G., Zhu, J.: Efficiently using prefix-trees in mining frequent itemsets. IEEE ICDM FIMI Workshop (2003).
9. Han, J., Kamber, M.: Data Mining: Concepts and Techniques, Morgan Kaufmann Publishing (2000).
10. Kochenberger, G., Glover, F.: Diversity data mining, Working Paper, The University of Mississipi (1999).
11. Orlando, S., Palmerimi, P., Perego, R.: Adaptive and resource-aware mining of frequent sets. IEEE Intl. Conf. on Data Mining (2002), 338–345.
12. Prais, M., Ribeiro, C. C.: Reactive GRASP: An application to a matrix decomposition problem in TDMA traffic assignment, INFORMS Journal on Computing **12** (2000), 164–176.
13. Resende, M. G. C., Ribeiro, C. C.: Greedy randomized adaptive search procedures, Handbook of Metaheuristics (F. Glover & G. Kochenberger eds.) (2003), 219–249.
14. Ribeiro, M. H., Trindade, V., Plastino, A., Martins, S.: Hybridization of GRASP metaheuristic with data mining techniques, Workshop on Hybrid Metaheuristics in conjunction with the 16th European Conf. on Artificial Intelligence (2004), 69–78.
15. Silva, G. C., Ochi, L. S., Martins, S. L.: Experimental comparison of greedy randomized adaptive search procedures for the maximum diversity problem, 3rd Intl. Workshop on Experimental and Efficient Algorithms, LNCS **3059** (2004), 498–512.
16. E. G. Talbi: A taxonomy of hybrid metaheuristics. Journal of Heuristics **8** (2002), 541–564.
17. Weitz, R., Lakshminarayanan, S.: An empirical comparison of heuristic methods for creating maximally diverse groups, J. of the Op. Res. Soc. **49** (1998), 635–646.

A New Multi-objective Particle Swarm Optimization Algorithm Using Clustering Applied to Automated Docking

Stefan Janson and Daniel Merkle

Department of Computer Science, University of Leipzig,
Augustusplatz 10-11, D-04109 Leipzig, Germany
{janson, merkle}@informatik.uni-leipzig.de

Abstract. In this paper we introduce the new hybrid Particle Swarm Optimization algorithm for multi-objective optimization ClustMPSO. We combined the PSO algorithm with clustering techniques to divide all particles into several subswarms. Strategies for updating the personal best position of a particle, for selection of the neighbourhood best and for swarm dominance are proposed. The algorithm is analyzed on both artificial optimization functions and on an important real world problem from biochemistry. The molecule docking problem is to predict the three dimensional structure and the affinity of a binding of a target receptor and a ligand. ClustMPSO clearly outperforms a well-known Lamarckian Genetic Algorithm for the problem.

1 Introduction

Particle swarm optimization is a population based metaheuristic especially useful for problems with a continuous, multi-dimensional search space. Real world problems often have the additional property that they have multiple objectives. If for a single-objective problem one is interested not only in the very best solution, but in several very good solutions, it may make sense to reformulate the single-objective problem as a multi-objective problem. A biochemistry problem of this kind is molecular docking. Molecular docking is successfully used in rational drug design, as it is a way to find potential drugs and is therefore of great industrial importance. It is widely accepted that drug activity is obtained by binding the drug molecule to a receptor macromolecule. Several approaches exist to attack the problem but still the approaches need immense computational effort to predict docked conformations [2,11]. Especially the class of docking simulation approaches is known to be computationally expensive. One important tool which uses this approach is Autodock [7]. Several random search algorithms are incorporated into Autodock 3.05. A Lamarckian Genetic Algorithm was shown empirically to perform best. In this report we will reformulate the molecular docking problem as a multi-objective optimization problem and use a hybrid of clustering and Particle Swarm Optimization (PSO) algorithm to successfully tackle the problem. Furthermore, the properties of the hybrid algorithm will be analyzed on artificial test functions.

M.J. Blesa et al. (Eds.): HM 2005, LNCS 3636, pp. 128–141, 2005.

In Section 2 we give a brief introduction to PSO and present ClustMPSO and its features. In Section 3 we will define the molecule docking problem and present how ClustMPSO can be used to successfully tackle the problem. Experiments on artificial functions and for the molecule docking problem are presented in Section 4. Conclusions are given in Section 5.

2   Particle Swarm Optimization

The original Particle Swarm Optimization algorithm was inspired by the behaviour of flocking birds [5]. A swarm of particles is searching for an optimum in a real-valued, multi-dimensional search space, given by an optimization function or provided by a specific application. Each particle is influenced by its own previous search experience and the information shared by other members of the swarm. The velocity of each particle is updated according to these points of attraction and the particle moves with the newly specified velocity.

A swarm consists of m particles and each particle $i \in \{1, \ldots, m\}$ is represented by its current position vector $x_i = (x_{i,1}, \ldots, x_{i,D})$ in the D-dimensional problem space and its velocity vector $v_i = (v_{i,1}, \ldots, v_{i,D})$. The best previous position of a particle is recorded as $y_i = (y_{i,1}, \ldots, y_{i,D})$. In every iteration particle i selects another particle n and uses its personal best position y_n and its own y_i to update its velocity. In the original PSO algorithm every particle of the swarm selected the global best particle (gbest) but the particles to choose from can also be restricted to a local neighbourhood (lbest) [4].

At each iteration of a PSO algorithm, after the evaluation of the objective function f, the personal best position of each particle i is updated, i.e., if $f(x_i) < f(y_i)$ then set $y_i = x_i$. The neighbourhood best particle n is selected and the new velocity of each particle i is determined in every dimension $d \in [1 : D]$ as given in (1).

$$v_{i,d} = w \cdot v_{i,d} + c_1 \cdot r_1 \cdot (y_{i,d} - x_{i,d}) + c_2 \cdot r_2 \cdot (y_{n,d} - x_{i,d}) \qquad (1)$$

Parameter w is called the inertia weight and it determines the influence of the old velocity.

The acceleration coefficients c_1 and c_2 are responsible for the influence of the local best position (cognitive aspect) and the neighbourhood best position (social aspect), respectively. The random values r_1 and r_2 are uniformly drawn from $[0, 1]$. After the velocity update the new position of particle i is then determined by $x_{i,d} = x_{i,d} + v_{i,d}$.

2.1   PSO for Multi-objective Optimization

The PSO algorithm has to be adjusted for the optimization of more than one objective. The multi-objective optimization function $f(x) : \mathbb{R}^D \rightarrow \mathbb{R}^O$ now returns a vector of O objective values. Hence, two solutions can no longer be compared directly. Instead, the concept of dominance is used to compare two solutions. A solution s_1 is said to dominate solution s_2 ($s_1 \prec s_2$) if for each component of s_1 and s_2: $s_{1,d} \leq s_{2,d}$ ($d = 1, \ldots, O$), with at least one strict inequality.

Several applications of PSO to multi-objective optimization have been proposed in literature, e.g. [3,12]. Most of these consider benchmark functions to test the quality of their approaches. Our ClustMPSO is sketched in Algorithm 1. and the details are described in the following.

ClustMPSO with Multiple Swarms. Our ClustMPSO approach uses multiple swarms instead of a single one. Each of these swarms S_k has its own non-dominated front $ND_k = \{i \in S_k \mid \nexists \, j \in S_k : j \prec i\} \subseteq S_k$. This subset of all the particles is used for selecting a neighbourhood best particle in equation (1). Particles within these non-dominated fronts ND_k can be dominated by particles within other swarms. Therefore, we are interested in the total non-dominated front ND of all particles of all swarms, $S = \bigcup_k S_k$. This front is defined accordingly, $ND = \{i \in S \mid \nexists \, j \in S : j \prec i\}$. For all of the operations within a swarm S_k only the respective non-dominated front ND_k is considered, the total non-dominated front ND is only required for determining whether a swarm is dominated (see later).

Update of the Personal Best Position. As in the single-objective PSO algorithm, a particle has to decide when to replace its personal best position y with its current position x. In the standard PSO algorithm the personal best position is replaced if $f(x) < f(y)$. But in the multi-objective case this decision is not as intuitive. Only replacing the personal best position if $f(x) \prec f(y)$ might be too restrictive for certain applications, because escaping a local minimum becomes difficult. Instead, also in such cases where $f(y) \nprec f(x)$ the personal best position can be updated. Possible approaches for this update are: 1.) This non-deteriorating operation is permitted with a certain probability. This approach was suggested in [12]. 2.) The update of the personal best is accepted, if the weighted sum of all components is smaller $\sum_d \lambda_d f(x)_d < \sum_d \lambda_d f(y)_d$ (recall that $f(x)$ and $f(y)$ are vectors with O components). The latter approach will be used in the real-world molecular docking problem, with $\lambda_d = 1$.

Neighbourhood Selection. Instead of one global best particle in each swarm S_k, a non-dominated front ND_k of particles is retained in which no two particles mutually dominate each other and all the remaining particles are dominated by at least one member of ND_k. Particles from the non-dominated front are selected as neighbourhood best particle n in the velocity update equation (1). Once a particle i has chosen its neighbourhood best particle n it sticks to this choice for $nSelect$ iterations. Only if the selected neighbourhood best particle n drops out of the non-dominated front, a new neighbourhood best particle is selected earlier.

Dominated Swarms. A swarm S_k is considered to be dominated if none of its particles is in the total non-dominated front ND. If a swarm is dominated for $maxDominated$ consecutive iterations, it is relocated. Therefore, a non-dominated swarm is randomly selected and each particle of the dominated swarm randomly selects the particle of the other swarm it is reset to; the personal best positions of the dominated particle is set to the personal best position of the selected particle.

2.2   Clustering

Clustering is used to separate the particles into several swarms. For this we use the K-means algorithm, which is a standard clustering method that provides good results quickly. K-means iteratively assigns all data vectors to K clusters $C_1, \ldots, C_K$ (often based on the Euclidian distance to the geometric centroid), and recomputes cluster centroids which are then used again for assigning the data vectors to clusters. This is repeated until there is no change in the assignment of the data points. For ClustMPSO the data vectors are the particles' best positions y_i or the vectors representing the objective values for the particles' best positions $f(y_i)$. This corresponds to clustering in search space or in objective space. All the particles within a cluster C_k form swarm S_k. Note, that clustering in objective space can only be effective for approaches with multiple swarms, if particles that are close in objective space are also close in search space. This is often not the case for real-world problems. The strategy would cluster particles from very different areas in the search space together, and the algorithm would be mislead. Nevertheless, for easy optimization functions this approach can be very effective. After a clustering step the non-dominated front of a swarm has to be updated. Recently, in [10] clustering in search space and objective space has been used for parallelization of multi-objective Evolutionary Algorithms. In [12] a combination of clustering and PSO was proposed. Particles of the non-dominated front are clustered in the search space for selecting the leader of subswarms. For this purpose the hierarchical single-connected clustering algorithm is used. A related approach has been presented for a multi-objective PSO algorithm in [8]. Although not clustering is used, the front is divided into several parts. The resulting subswarms are used to cover the Pareto-optimal front with high granularity.

3   Molecular Docking

Given a biological target receptor (a protein or DNA) and a flexible ligand (a small molecule) the molecule docking problem is to predict the three dimensional structure and the affinity of a binding. For a broad overview of different molecular docking approaches see [2]. Here we will focus on docking simulation methods. A ligand begins at a random position and explores the docking space until the best conformation is found. From an optimization point of view such a complex of protein and ligand has a scoring function which has to be minimized.

3.1   Theory

As scoring function an empirical binding free energy function can be used to successfully rank complexes. Let r_{ij} be the internuclear separation of two atoms.

Algorithm 1. Generic Pseudo code of ClustMPSO

Swarm S of m particles

Initialize location $x_i = (x_{i1}, \ldots, x_{iD})$ and velocity $v_i = (v_{i1}, \ldots, v_{iD})$ of each particle $i \in S$. Best position $y_i = x_i$.

Initialize centroid vector of cluster C_k to y_i for a randomly selected particle $i \in S$

repeat

  repeat

    cluster the m particles into K different swarms. Each cluster C_k **(1)**

    representing one swarm S_k, $k = 1, \ldots, K$.

  until required cluster quality is met

  for each Swarm S_k **do**

    for each particle $i \in S_k$ **do**

      evaluate objective function $f(x_i)$ at the particle's location x_i

      update the personal best position y_i **(2)**

    end for

    update non-dominated front $ND_k = \{i \in S_k \mid \nexists\, j \in S_k : j \prec i\}$

    for each particle $i \in S_k$ **do**

      if n has been selected $nSelect$ iterations ago **or** $n \notin ND_k$ **then**

        select a neighbourhood best particle n from ND_k **(3)**

      end if

      update the velocity v_i in each dimension d:

        $v_{i,d} = w \cdot v_{i,d} + c_1 \cdot r_1 \cdot (y_{i,d} - x_{i,d}) + c_2 \cdot r_2 \cdot (y_{n,d} - x_{i,d})$

      compute the new location of the particle: $x_{i,d} = x_{i,d} + v_{i,d}$

    end for

  end for

  update total non-dominated front $ND = \{i \in S \mid \nexists\, j \in S : y_j \prec y_i\}$, $S = \bigcup S_k$

  for each Swarm S_k **do**

    if a particle of S_k is in ND **then**

      $S_k.dominatedCount = 0$

    else

      $S_k.dominatedCount + +$

    end if

  end for

  for each Swarm S_k **do**

    if $S_k.dominatedCount \geq maxDominated$ **then**

      relocate the swarm S_k **(4)**

    end if

  end for

until stopping criterion is met

$$\Delta G = \Delta G_{vdw} \sum_{i,j} \left(\frac{A_{ij}}{r_{ij}^{12}} - \frac{B_{ij}}{r_{ij}^{6}} \right) + \Delta G_{hbond} \sum_{i,j} E(t) \left(\frac{C_{ij}}{r_{ij}^{12}} - \frac{D_{ij}}{r_{ij}^{10}} + E_{hbond} \right)$$

$$+ \Delta G_{elec} \sum_{i,j} \frac{q_i q_j}{\varepsilon(r_{ij}) r_{ij}} + \Delta G_{tor} N_{tor} + \Delta G_{sol} \sum_{i_C, j} S_i V_j e^{(-r_{ij}^2 / 2\sigma^2)}$$

Fig. 1. Free Energy Function used in Autodock 3.05

Then the suggested empirical binding free energy function as used in Autodock 3.05 is shown in Figure 3.1. The ΔG_* constants in the above equation are empirically determined by a regression from a set of 30 complexes with known binding constants. The first three terms in the equation are terms for molecular dynamics which measure the in vacuo contributions of the binding free energy. The first term estimates the van der Waals potential energy with typical 12-6-Lennard-Jones parameters. The second term models the energy for hydrogen bonds using a 12-10 hydrogen bonding term and a directional weight $E(t)$ based on the angle t between the probe and the target atom. E_{hbond} denotes the estimated average energy of hydrogen bonding of water with a polar atom. The third term models the electrostatic energy between two atoms with charges q_i and q_j. The fourth term models the restriction of internal rotors and global rotation and translation. As it models the loss of torsional degrees of freedom upon binding it is proportional to the number of sp^3 bonds N_{tor}. Note, that this term is constant and can be excluded during optimization and added in a post processing step to calculate the predicted energy of the binding. The fifth term models the desolvation term. For each carbon atom of the ligand fragmental volumes of surrounding proteins are weighted with an exponential function. This sum is weighted with the atomic solvation parameter to give the final estimated desolvation energy. Variants of the fifth term were deeply investigated in [7]. To rapidly evaluate the energy terms Autodock 3.05 uses trilinear interpolation. For this, grid maps for each type of atom presented in the ligand are needed, where each point in the grid stores the potential energy of a probe atom at that position. In addition to these affinity grid maps, an electrostatic grid map is needed which stores the Coulombic interactions between the macromolecule and a probe of fixed charge. For a much more detailed description and investigation of the free energy function and variants of it and for tables for the used constants, see [7].

In the further sections ΔG_{inter} denotes the intermolecular interaction energy between the protein and the ligand, whereas ΔG_{intra} denotes the intramolecular interaction energy of the ligand only. For the calculation of G_{intra} all pairs of atoms in the ligand are used, which have at least three bonds in between. Energy terms that exclude the torsional free energy are called docked energies. The docked energy term used for optimization is $E = \Delta G_{inter} + \Delta G_{intra}$. Note, that in Autodock 3.05 all possible optimization algorithms aim to minimize E in a single-objective fashion.

3.2   ClustMPSO for Molecular Docking

Usually, when searching for bindings with software tools like Autodock 3.05, several runs are performed on the same instance. The reason for this is that from a biochemical point of view it makes sense not only to find the very best docking, but also interesting dockings which still have small (but maybe not the smallest) docking energies. In such dockings the position of the ligand may be quite different from the optimal solution. In this paper we propose a bi-objective approach. The two objectives used are the intramolecular interaction energy ΔG_{intra} and the intermolecular interaction energy ΔG_{inter}. In contrast to other real-world problems the sum of different objectives makes sense for the molecular docking problem. Hence, the docking energy $\Delta G_{intra} + \Delta G_{inter}$ will be used when updating the personal best position of a particle and for comparing results of algorithm ClustMPSO with the single objective optimization methods of Autodock.

Same as in the Lamarckian Genetic Algorithm (LGA) and Simulated Annealing (SA) approach of Autodock, the number of degrees of freedom of the optimization problem is $7 + N_{tor}$, where N_{tor} is the number of rotatable bonds of the ligand. The position of the ligand is defined by a three-dimensional vector, the orientation is defined by a quaternion. Note that for the clustering step of algorithm ClustMPSO only the position of the ligand is used.

4   Experiments

In this section we will at first present results on a well-known artificial function and then apply ClustMPSO to the molecular docking problem. For all our experiments we used the common PSO parameter $w = 0.729$ and $c_1 = c_2 = 1.494$. The remaining ClustMPSO parameters and the specific changes to the algorithm for each application are described later.

4.1   Artificial Functions

We applied our ClustMPSO algorithm to several standard multi-objective benchmark functions, where we want to achieve a large number of non-dominated solutions quickly. Therefore, a non-dominated particle behaves differently from a dominated one. It only accepts its current position x as new personal best position y if x dominates y. A dominated particle also accepts x if $x \prec y$ and with probability 0.5 if $y \not\prec x$, see Algorithm 2.. Also, a non-dominated particle selects itself as neighbourhood best particle n in the velocity update equation (1). The swarmsize used is $m = 256$ and the number of clusters is varied in $K \in \{1, 3, 16, 32\}$. The other parameters are $nCount = 10$ and $maxDominated = 10$. Note, that due to the limitation of space we only present results on one test function in more detail and will then focus on the molecule docking.

The artificial test function T has been proposed in [?]. $(f_1(x), f_2(x))$ has to be minimized. The search dimension is $D = 3$ and $x_i \in [-5; 5]$.

$$f_1(x) = \sum_{i=1}^{D-1} - 10 \cdot e^{-0.2\sqrt{x_i^2 + x_{i+1}^2}}$$
$$f_2(x) = \sum_{i=1}^{D} \left(|x_i|^{0.8} + 5\sin(x_i^3) \right)$$

The pareto front of test function T consists of three sections that correspond to three different areas in the search space. We compared the two clustering strategies of clustering in objective (clust-objective) or in search space (clust-search). In Figure 2 the non-dominated solutions for $K = 1$ and $K = 3$ clusters

Algorithm 2. ClustMPSO Adaptions for Artificial Optimization Functions

if $f(x_i) \prec f(y_i)$ or $\left(f(y_i) \not\prec f(x_i) \text{ and } \mathrm{Rand}(0,1) \le 0.5\right)$ then
   $y_i = x_i$ (2)
end if

if $i \in ND_k$ then
   $n = i$
else   (3)
   randomly select n from ND_k
end if

reset each particle i from S_k to randomly selected particle j from randomly   (4)
selected non-dominated swarm S_y: $y_i = y_j$

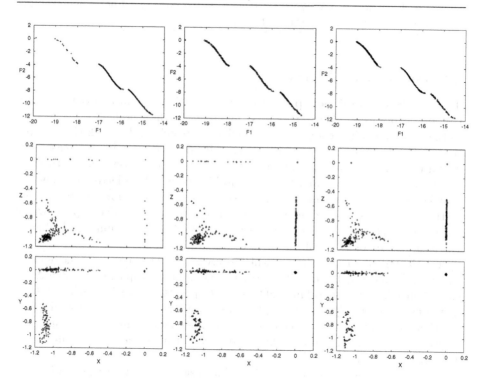

Fig. 2. Results of ClustMPSO on test function T; shown are the objective space (top row) and the search space (middle and bottom row) after 500 steps; $K = 1$, $K = 3$ clustering in search space, $K = 3$ clustering in objective space (from left to right).

are displayed. The total number of non-dominated solutions is 229 for $K = 1$, 178 for $K = 3$ when clustering is performed in search space (left and middle column), and 228 for $K = 3$ if clustering is done in objective space. It can be clearly seen, that the distribution of the particles is much worse when using only one swarm. The reason is that using only one swarm decreases diversity, and only very few particles are located in the area ($x \approx 0, y \approx 0, z$), which corresponds to the first section of the non-dominated front (comp. second and third row of

Figure 2). Note also, that clustering in objective space clearly divides the non-dominated front in its three parts - this is due to the relatively easy structure of the artificial test function (when compared to real world applications).

Algorithm 3. ClustMPSO Adaptions for Autodock

if $f(x_i) \prec f(y_i)$ **or** $(f(y_i) \not\prec f(x_i)$ **and** $\sum_d f(x_i)_d < \sum_d f(y_i)_d)$ **then**
$\quad y_i = x_i$
end if

$\hfill (2)$

---

randomly select n from ND_k $\hfill (3)$

---

reset each particle i from S_k to randomly selected particle j from randomly selected non-dominated swarm S_y: $y_i = y_j$ $\hfill (4)$

4.2 Docking Experiments

Test Instances and Setup. For the investigation of ClustMPSO for automated docking we used two instances. The name of the instances are according to their corresponding code of the Brookhaven Protein Data Bank (PDB-code) [1]. For test instance 1hvr the protein used for docking is HIV-1 protease. The ligand is the HIV-1 protease inhibitor XK-263 that can prevent the maturation of virions of HIV. The ligand has 10 rotatable bonds, which leads to 17 degrees of freedom. For test instance 4cha the large molecule used for docking is chymotrypsin, a digestive enzyme that can perform proteolysis. The peptide which has to be docked to chymotrypsin is Benzoyl-Ala-Asp(OGp)-amide. The ligand has 9 rotatable bonds, resulting in 16 degrees of freedom.

For the trilinear interpolation, which is used for the evaluation of the docking energy, grid maps with $101 \times 101 \times 101$ points and a grid spacing of 0.375 Å are used. Note, that in [7] a grid map of $61 \times 61 \times 61$ points and a grid spacing of 0.375 Å was used. Obviously, expanding the grid map increases the possible docking positions and the optimization problem becomes more difficult. The size and position of both grid maps and the target receptor molecule is shown in Figure 3 for instance 1hvr.

The ClustMPSO algorithm is adjusted as described in Algorithm 3.. For the update of the personal best position we used the criterion based on the sum of the components of the objective function, i.e., a new personal best position is accepted, if $\sum_d f(x)_d < \sum_d f(y)_d$ (see Section 2). Each particle (dominated and non-dominated) selects its neighbourhood best particle n for the velocity update equation (1) randomly from the non-dominated front ND_k of the respective swarm S_k. We used a total of $m = 512$ particles. The number of clusters is varied in $K \in \{1, 2, 4, 8, 16, 32, 64, 128\}$ and the maximum number of iterations a swarm can be dominated is $maxDominated \in \{1, 10, 50, 100, 200, 400, \infty\}$. When varying the number of clusters $maxDominated$ is 200 and when varying $maxDominated$ we set $K = 32$. The duration of a selection of a specific neighbourhood best particle is $nCount = 10$.

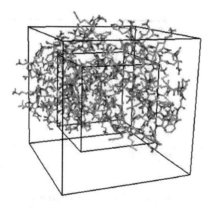

Fig. 3. Molecule: Human Immunodeficiency Virus Type 1 (HIV-1) protease; large cube: position and size of the grid map ($101 \times 101 \times 101$ points with a grid spacing of 0.375 Å) used for our investigations; small cube: grid map used in [7] ($61 \times 61 \times 61$ points with a grid spacing of 0.375 Å).

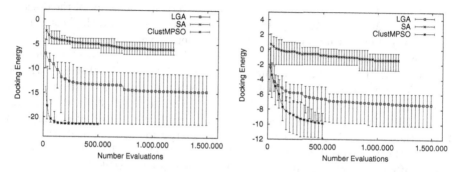

Fig. 4. Convergence behavior of SA, LGA, and ClustMPSO ($K = 32$, $maxDominated = 200$): shown are the worst, average, and best found docking energy (10 runs); left: instance 1hvr, right: instance 4cha.

We compared our results with Simulated Annealing dockings and Lamarckian Genetic Algorithm dockings, which were performed with Autodock 3.05. The 1hvr instance is included in the Autodock 3.05 distribution and uses the following parameters, which were also used in [7]. Ten runs were performed for both SA and LGA. The SA approach used 50 cycles and a maximum of 25.000 accepted or 25.000 rejected steps, whichever came first (the total number of evaluations is at least 1.2 million). For the LGA the number of generations was 27.000 with a population size of 50. In 6% of all evaluations a genome underwent a Lamarckian local search with 300 iterations. The LGA was stopped after 1.5 million evaluations. The local search was based on the Solis Wets local [9] search. For further details of the SA and GA parameters see [7] and the Autodock 3.05 example instances.

Results. Before illuminating the multi-objective behavior of ClustMPSO for automated docking, we compare algorithm ClustMPSO with the single-objective

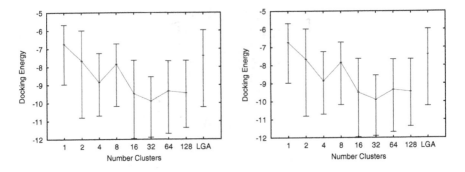

Fig. 5. Worst, average, and best found docking energy for ClustMPSO after 512.000 evaluations (10 runs) with a maximal number of clusters $K \in \{1, 2, 4, 8, 16, 32, 64, 128\}$; worst, average, and best found docking energy for LGA after 1.5 million evaluations; left: instance 1hvr, right: instance 4cha.

algorithms SA and LGA with respect to the convergence behavior of the docking energy $E = \Delta G_{inter} + \Delta G_{intra}$. In Figure 4 the average energy, the best energy, and the worst energy found in 10 runs after t energy evaluations is shown. It can be seen that ClustMPSO clearly outperforms both other approaches. On test instance 1hvr the average performance of ClustMPSO after 1000 steps (512.000 evaluations) was -21.389 (kcal mol^{-1}) which is better than the best result that was found with LGA after 10×1.5 million evaluations (-21.384). Also on instance 4cha ClustMPSO achieves much better results with less evaluations than both other approaches. The lowest docking energy found by ClustMPSO was -11.888, the best result found by the LGA was -10.219. The average docking energy after 512.000 evaluations was -9.887 for the ClustMPSO approach and -7.395 for the LGA approach. The influence of the maximal number of clus-

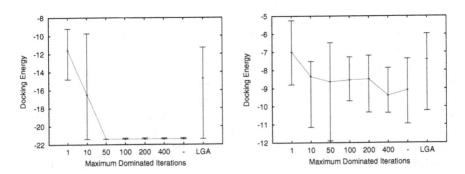

Fig. 6. Worst, average, and best found docking energy for ClustMPSO after 512.000 evaluations (10 runs) for different number of maximum dominated iterations $maxDominated \in \{1, 10, 50, 100, 200, 400, \infty\}$ clusters; worst, average, and best found docking energy for LGA after 1.5 million evaluations; left: instance 1hvr, right: instance 4cha.

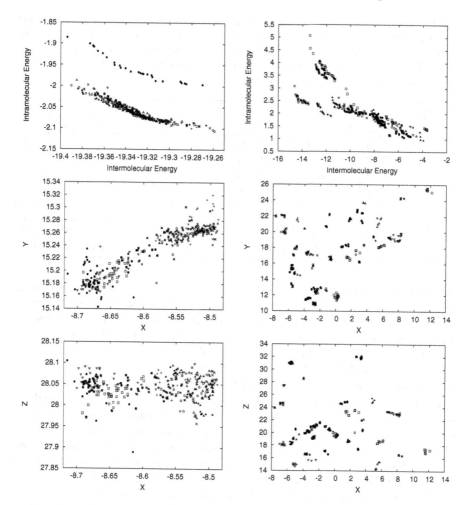

Fig. 7. Positions of all non-dominated particles in objective space (top row) and three-dimensional docking space (middle and bottom row) for instances 1hvr (left) and 4cha (right) for all 10 runs of ClustMPSO after 512.000 evaluations; paramters used: $K = 32, maxDominated = 200$; different symbols represent different test runs.

ters used for ClustMPSO is depicted in Figure 5 and 6. As the fitness landscape in the autodocking instances has a much more complex structure than the artificial functions, ClustMPSO can clearly benefit from a higher maximal number of clusters K. For $K \in \{8, 16, 32, 64\}$ the worst result of all runs was -21.297 for instance 1hvr. For a small K (1,2, and 4) the solutions are significantly worse. Also in instance 4cha the best average docking energies when using $K \in \{1, 2, 4, 8\}$ clusters are worse then the worst average docking energies when using $K \in \{16, 32, 64, 128\}$.

The influence of the maximal number of iterations $maxDominated$, that a swarm is allowed to be dominated is shown in Figure 6. The influence of

Table 1. Sizes of the swarms and number of non-dominated particles for instances 1hvr and 4cha after 1000 steps of algorithm ClustMPSO; K: maximal number of clusters, $min_k\{|S_k|\}$: minimal number of particles in a swarm, $max_k\{|S_k|\}$: maxmimal number of particles in a swarm, *act*: number of active swarms with ≥ 1 particles, *nd*: number of non-dominated particles; results are averaged over 10 runs.

	1hvr				4cha			
K	$min\{|S_k|\}$	$max\{|S_k|\}$	*act*	*nd*	$min\{|S_k|\}$	$max\{|S_k|\}$	*act*	*nd*
1	512	512	1	24.7	512	512	1	15
2	171.2	340.8	2	37.9	215.7	296.3	2	15.2
4	66.2	202.9	4	36.1	55.7	210.6	4	21.2
8	33.8	104.7	7.9	40.7	27.8	107.9	8	23.1
16	11.2	65.3	15.9	40.1	8.9	65.9	16	29.7
32	4.4	34	30.8	43.2	3.3	40.5	31.4	37.1
64	1.7	20.2	62.1	53.7	1.3	25.1	61.5	58.4
128	1	15.4	119.9	76.4	1	15.3	120.5	76.1

the parameter can be tremendous. In both instances the worst average docking energies were found for the case, when dominated swarms are replaced very often (every iteration or every 10 iterations). Although very good results were achieved for the case $maxDominated = \infty$ (i.e., a swarm never gets replaced), it should be noted that increasing the parameter may decrease exploration.

The results with respect to multiple objectives are shown in Figure 7. It can be seen, that instance 1hvr is comparatively simple, as there is basically only one important docking position from a biochemical point of view (note the scale of the figures in the left column). ClustMPSO also found other solutions which were non-dominated, but where the docking energy is much larger. These solutions ($E > -21.0$) are not shown in Figure 7 as they would disproportionally distort the non-dominated front in the most important area. Instance 4cha is much more interesting as several different docking positions were found with ClustMPSO as can be seen in the position of the solutions in the docking space.

The sizes of the swarms and the number of non-dominated particles for different numbers of maximal clusters K is given in Table 1. Each number is averaged over 10 runs with 1000 steps for each run. It can be clearly seen that on both test instances the number of particles in the non-dominated front increases for a larger number of maximal clusters. Note, that too many clusters can decrease the overall performance (comp. Figure 5). Furthermore, if K is too large, the number of active swarms (i.e., swarms with ≥ 1 particles) decreases. While for $K = 4$ all 4 swarms are active in all runs, the number of active swarms decreases to 93.9% for $K = 128$. Hence, using maximal 32 or 64 clusters is a good tradeoff between diversity and good performance.

5   Conclusion

In this paper we presented the new hybrid Particle Swarm Optimization algorithm for multi-objective optimization ClustMPSO. Within one iteration all particles are clustered to separate all particles into multiple swarms. Several variants to guide the particles and to replace them were tested with respect to multi-objective optimization. ClustMPSO was applied to a real-world problem from biochemistry. The molecule docking problem is to predict the three-dimensional structure of a binding of a target receptor and a ligand. We showed the ClustMPSO was able to clearly outperform a well-known and successful Lamarckian genetic algorithm.

References

1. F.C. Bernstein, T.F. Koetzle, G.J.B. Williams, E.F. Meyer, M.D. Brice, J.R. Rogers, O. Kennard, T. Shimanouchi, and M. Tasumi. The protein data bank: a computer-based archival file for macromolecular structures. J. Mol. Biol., 112:535-542, 1977.
2. I. Halperin, B. Ma, H. Wolfson, and R. Nussinov. Principles of docking: An overview of search algorithms and a guide to scoring functions. Proteins: Structure, Function, and Genetics, 47(4):409-443, 2002.
3. X. Hu and Russell Eberhart. Multiobjective optimization using dynamic neighborhood particle swarm optimization. In D.B. Fogel et al., editor, Proc. of the CEC2002, pages 1677-1681. IEEE Press, 2002.
4. J. Kennedy. Small worlds and mega-minds: Eects of neighborhood topology on particle swarm performance. In P.J. Angeline et al., editor, Proc. of the Congress on Evolutionary Computation, volume 3, pages 1931-1938. IEEE Press, 1999.
5. J. Kennedy and R. Eberhart. Particle swarm optimization. In IEEE International Conference on Neural Networks (ICNN'95), volume 4, pages 1942-1947. IEEE Press, 1995.
6. F. Kursawe. A Variant of Evolution Strategies for Vector Optimization. In H. P. Schwefel and R. MÄanner, editors, Proc. 1st Workshop Parallel Problem Solving from Nature (PPSN), volume 496 of LNCS, pages 193-197, Berlin, 1991. Springer.
7. G.M. Morris, D.S. Goodsell, R.S. Halliday, R.Huey, W.E. Hart, R.K. Belew, and A.J. Olson. Automated docking using a lamarckian genetic algorithm and an empirical binding free energy function. Journal of Computational Chemistry, 19(14):1639-1662, 1998.
8. S. Mostaghim and J. Teich. Covering pareto-optimal fronts by subswarms in multi-objective particle swarm optimization. In Proc. of the Congress on Evolutionary Computation (CEC '04), pages 1404-1411, 2004.
9. R. Solis and T. Wets. Minimization by random search techniques. Mathematics of Operations Research, 6:19-30, 1981.
10. F. Streichert, H. Ulmer, and A. Zell. Parallelization of multi-objective evolutionary algorithms using clustering algorithms. In Proc. Conf. on Evolutionary Multi-Criterion Optimization (EMO 2005), volume 3410 of LNCS, pages 92-107, 2005.
11. M. Teodoro, G.N.Jr. Phillips, and L.E. Kavraki. Molecular docking: A problem with thousands of degrees of freedom. In Proc. of the IEEE International Conference on Robotics and Automation (ICRA), pages 960-966. IEEE press, 2001.
12. G. Toscano Pulido and C.A. Coello Coello. Using Clustering Techniques to Improve the Performance of a Particle Swarm Optimizer. In K. Deb et al., editor, Proc. of the Genetic and Evolutionary Computation Conference (GECCO 2004). Part I, number 3102 in LNCS, pages 225-237, Berlin, June 2004. Springer.

A Hybrid *GRASP-Path Relinking* Algorithm for the Capacitated *p – hub* Median Problem

Melquíades Pérez, Francisco Almeida, and J. Marcos Moreno-Vega

Escuela Técnica Superior de Ingeniería Informática,
Universidad de La Laguna
Address: DEIOC
Avda. Astrofísico Francisco Sánchez,
s/n 38271, La Laguna, Spain
Phone: -34(922) 318173, Fax: -34(922) 318170
{melperez, falmeida, jmmoreno}@ull.es

Abstract. The $p – hub$ median problem is an *NP* hard location – allocation problem, that consists of finding p points to establish facilities and the assignment of the users to these points. In the capacitated version of this problem, each *hub* has got a maximum capacity limiting the traffic to be assigned. A new evolutionary approach that has been very effective for solving optimization problems is *Path Relinking*, an extension of Scatter Search that links solutions over neighborhood spaces. *GRASP* is a well-known randomized multistart metaheuristic. In this paper, we present a hybrid *GRASP-Path Relinking* for the capacitated $p – hub$ median problem where the *GRASP* is used to construct the population of the *Path Relinking*. Computational results demonstrate that the hybrid *GRASP-Path Relinking* provides better solutions, in terms of both running times and solution quality.

1 Introduction

There are many real situations where several nodes must interact among each other by sending and receiving traffic flow of some nature. The flow exchanged may represent data, passengers, merchandises, express packages, etc. Generally, in these situations is useful to find an optimal location of several points called *hubs*. The hubs act like exchanging points. The objective is to find the location of the *hubs* and to allocate the non hub nodes to the *hubs* minimizing an objective function that describes the interchanged flow and its cost. The *hubs* are fully interconnected and the traffic between any pair of nodes is routed through the hubs.

This problem was first formulated, by O'kelly[11], as an integer quadratic problem. After that, different variants of the problem have been formulated. For example, we may consider single or multiple assignments, where nodes can be assigned to one or several hubs respectively. The hubs may have constraints on the amount of traffic (capacity). The hubs may have some set up cost associated to the hubs, etc.

We will approach here the capacitated single allocation *p-hub* median problem. We consider a variant to the problem formulated in [14], where the hubs have been estab-

M.J. Blesa et al. (Eds.): HM 2005, LNCS 3636, pp. 142–153, 2005.

lished in advance and the set up cost is *zero*. In this paper we propose the application of two heuristics methods to the capacitated *p-hub* median problem.

This problem can be modelized as a location – allocation problem:

Let N be a set of n demand points. We denote, for any pair of nodes i and j:

$$W_{ij} = \text{number of units sent from } i \text{ to } j$$
$$C_{ij} = \text{standard cost per unit of flow from } i \text{ to } j$$

Normally, $W_{ii} = 0$ and $C_{ii} = 0$, for all i, although there are some cases with $W_{ii} \neq 0$ and/or $C_{ii} \neq 0$. If one of the points, i or j, is a hub, the standard cost per unit C_{ij} is assumed. If they are both hubs, then the standard cost per unit of traffic is normally reduced and it is equal to αC_{ij}, where α is a parameter. In general, $\alpha \leq 1$ to reflect the effect of the reduced cost in inter–hubs flows. We can also consider parameters δ and γ to capture the influence of reduced or incremented costs for non hub points. There are not constraints for the parameters δ and γ.

Let X_{ij} and Y_j be the decision variables defined as:

$$X_{ij} = \begin{cases} 1 & \text{if point } i \text{ is allocated to hub } j \\ 0 & \text{otherwise} \end{cases} \qquad Y_j = \begin{cases} 1 & \text{if point } j \text{ is a hub} \\ 0 & \text{otherwise} \end{cases}$$

Let the variables:

$$O_i = \sum_{j \in N} W_{ij} \text{ , denotes the total flow with origin at node } i.$$

$$D_i = \sum_{j \in N} W_{ji} \text{ , denotes the total flow with destiny at node } i.$$

Also, let ξ_k be the capacity of each hub k.

Then, the capacitated single allocation *p-hub* median location problem can be formulated as:

$$minf(x) = \sum_i \sum_j W_{ij} \times \left(\delta \sum_k X_{ik} C_{ik} + \gamma \sum_m X_{jm} C_{jm} + \alpha \sum_k \sum_m X_{ik} X_{jm} C_{km} \right)$$

subject to :

$$X_{ij} \leq Y_j \qquad \qquad for \; i = 1,...,n \; and \; j = 1,...,n \qquad (1)$$

$$\sum_{i \in N} X_{ij} = 1 \qquad \qquad for \; i = 1,...,n \qquad (2)$$

$$\sum_{j \in N} Y_j = p \qquad \qquad (3)$$

$$\sum_{i \in N} (O_i + D_i) X_{ij} \leq \xi_j \cdot Y_j \qquad for \; j = 1,...,n \qquad (4)$$

$$X_{ij}, Y_j \in \{0, 1\} \qquad \qquad for \; i = 1,...,n \; and \; j = 1,...,n \qquad (5)$$

Constraint (1) ensures that a non hub point is allocated to a location j if a hub is located in that site. Constraint (2) guarantees that each point is served by one and only one hub. Constraint (3) generates the correct number of hubs. Constraint (4) guarantees that the capacity of each hub is not exceeded. And finally, the classical binary constraint (5).

Capacitated models for hubs problems have demonstrated to be a better approach to many realistic situations than the classical single versions. The capacitated single allocation *p-hub* median problem is also an *NP*–hard problem. Furthermore, even if the hubs are located, the problem of allocating points is also *NP*–hard. Therefore, it is necessary to have recourse to heuristic solution approaches, especially when the problem contains a very large number of nodes.

In this paper, we present a new *Path Relinking* approach to solve the single allocation capacitated *p–hub* median problem. This new proposal uses systematic neighborhood-based strategies to explore the feasible region of the search space and provides good results even for large problem sizes. Furthermore, we show a hybrid algorithm that combines *GRASP* and *Path Relinking*, using *GRASP* as the preprocessing method for the *Path Relinking*. The paper has been structured as follows. In section 2 we present the state of the art of the algorithms proposed for the *p − hub* median problems. Section 3 introduces the principles of the *Path Relinking* method. Section 4 describes the implementation of the *Path Relinking* algorithm that we apply to solve the problem. Section 5 consists of a brief explanation of the *GRASP* metaheuristic and its basic elements. Section 6 expounds the combination of *GRASP* and *Path Relinking*. Finally, Section 7 reports computational results analyzing advantages and disadvantages of the hybrid approach. We conclude that the standard *Path Relinking* is widely benefited from the hybridization.

2   The State of the Art

Several heuristics have been proposed for the p–hub problem. They can be classified attending to the phase (location or assignment) they develop first. Some of them are focused on the study of the locational part and others on the assignment part. The first method to solve the *p-hub* problem was proposed by O'Kelly [11], based on an exhaustive search with two different assignment methods, HEUR1 and HEUR2. This work introduces a set of problem instances that have been used in the literature as test problems (CAB). Klincewicz [7] proposes more sophisticated heuristics. Foremost, he considers the traffic among the hubs in addition to the distance criteria. These two criteria are combined into a multicriteria assignment procedure. Klincewicz also considers exchange heuristics, taking into account only the configurations that yield immediately promising improvements, and proposes an associated *clustering* heuristic. Klincewicz [8] also investigates Tabu Search and *GRASP*. The heuristics proposed by Klincewicz provide running times better than the running times obtained with O'Kelly's proposals. A version of Tabu Search, called TABUHUB, is suggested in [12]. This heuristic uses a tabu procedure for the location and allocation phases. The computational experience developed shows that this last procedure is better than the O'Kelly's heuristics, HEUR1 and HEUR2. TABUHUB does not run as fast as Klincewicz's heuristics, but produces higher quality solutions and constitutes one of the best heuristics known to solve the *p–hub* median problem. M. Pérez et al. [10] present a hybrid method that combines a genetic algorithm with multistart search. The main contribution is a new codification of the solutions and an associated selection of the operators, to produce a computational behaviour that is sometimes better than TABUHUB. Krishnamoorthy et al. [1] present new formulations reducing the number of variables and constraints of the problem. They develop a heuristic algorithm based on Simulated Annealing and use the upper bound provided by this heuristic to create a

Branch & Bound solution method. Computational results for the CAB problems and for a new set of problems (called the AP set) are presented. No comparisons with another heuristic approaches are available.

As stated before, single allocation *p-hub* median problem have been extensively studied, however, the literature about capacitated *p-hub* problems is scarce. In particular, the version without establishing costs that we consider here has not been solved at all. Krishnamoorthy et al. [14] proposed the problem with fixing costs. They present a MILP formulation with fewer variables and constraints than the previous proposals and suggest two simple heuristics based on simulated annealing and random descent. Aykin [15], investigates about the capacitated multiple allocation problem. At the same time, Aykin considers direct connections between origins an destinations.

Some of the heuristics mentioned above display a good behaviour in particular cases, however, when the size of the problem increases, the performance of the heuristic decreases. Nowadays, the development of fast and efficient heuristics to solve large instances of the *p-hub* problem remains a challenging problem. We present, in this paper, an adaptation of the *Path Relinking* metaheuristic to this problem. This new heuristic introduces a remarkable improvement in the size of the problem solved while maintaining reasonable running times. Besides, we have developed an hybrid heuristic combining *GRASP* and *Path Relinking*. This hybrid method improves the results of the simple *Path Relinking* algorithm, both in running times and in the quality of the solutions.

3 *Path Relinking* Principles

The *Path Relinking* (*PR*) method is an evolutionary metaheuristic procedure founded on the Scatter Search design originally formulated in Glover [4] in the 70s, but replacing the Euclidean spaces used in Scatter Search with neighbourhood spaces as a basis for combining solutions (Glover [5], Glover, Laguna and Marti [6]). Nowadays the literature contains many different applications of the procedure to hard optimization problems (see, e.g., the survey [6]). The method operates on a set of solutions called the *Reference Set*, *RS*, using strategies to select, combine, improve and generate new and better solutions from this set.

The Reference Set is a collection of the |RS| more representative solutions of the problem, selected from an initial set of trial solutions, *P*.

PR applies the mechanisms of diversification and intensification to update the Reference Set and to guide the heuristic towards an optimal solution of the problem. The diversification of *PR* is customarily applied by updating a subset of the Reference Set with solutions not in the Reference Set that maximize the distance to the closest member of Reference Set. The remaining solutions of the Reference Set are updated through the intensification scheme. Usually, the intensification is executed by combining different solutions to create new initial solutions for the local search procedures.

PR uses a mechanism for combining solutions that creates one or more neighbourhoods for moving from one solution to another, in the manner of neighbourhood search. Then *PR* executes a sequence of moves in the selected neighbourhood(s), to generate intermediate solutions that constitute the combinations produced (by the *PR* interpretation of the meaning of "combination"). In this way, the neighbourhood structure characteristically produces new solutions that contain some mix of the attributes of the parents, while preserving desirable properties. Multiple parents can be used simultaneously by constructing a neighbourhood path, selecting an appropriate evaluation criterion.

The success of *PR* is based on a good integration of the methods to select, combine and improve the solutions that update the Reference Set at each step of the procedure. The selection is carried out by partitioning the Reference Set into components. Here we use a 2-component partitioning [6] that divides the Reference Set in two subsets RS_1 and RS_2, respectively ($|RS_1| + |RS_2| = |RS|$). The first subset is composed of the $|RS_1|$ best solutions. The $|RS_2|$ most diversified solutions are included on the second subset. Then, two types of selection are performed. The solutions obtained by the combination and improvement are selected to update the first subset. The second subset is updated with the solutions in *P* more distant from the Reference Set.

The determination of the solutions to combine within the Reference Set is performed by generating subsets of the Reference Set. The subsets can be generated, for example, as follows: All the subsets of 2 solutions, all the subsets of 3 solutions derived from the 2-solutions subsets, adding the best solution, all the subsets of *i* solutions derived form the (i-1)-solutions subset, adding the best solution (*i* = 4 to *RS*), etc.

After that, the solutions obtained with the combination are improved with a simple local search.

As stated in [16] "*PR* can be considered an extension of the Combination Method of Scatter Search. Instead of directly producing a new solution when combining two or more original solutions, *PR* generates paths between and beyond the selected solutions in the neighbourhood space". On the other hand, the Combination Method of Scatter Search produces solutions that belong to some path connecting the initial solutions. For these reasons, the differences between *PR* and Scatter Search are quite subtle. In fact, some authors may consider that a given implementation corresponds to a Scatter Search and others could consider that tally with a *PR*.

4   *Path Relinking* for the Capacitated Single Allocation *p-hub* Median Problem

4.1   Codification of Solutions

We present our representation for the solutions of the *p–hub* problem and its utilization within the *PR* approach. The matrix with restrictions (1), (2), (3) and (4) is a natural representation of a solution. Nevertheless, this representation is difficult to use when designing efficient combination methods. Then, we propose the following representation:

Each solution is represented by an array of size *n* that indicates the hubs and the assignment of the rest of nodes to these hubs. Lets assume that the set of demand points is indexed by the set $L=\{1, \dots, n\}$, the idea of our representation is the following: the first *p* positions of the array, called *location–array*, are the ordered indexes of the hubs; and the last $(n – p)$ positions, called *allocation–array*, designate the allocations of the non hub nodes, assuming that the hubs are allocated to themselves.

Suppose, for example, that $n = 5$, $p = 2$, and $L = (1\ 2\ 3\ 4\ 5)$. In this context, the solution X, where hubs are the nodes 2 and 4, and the first and the fifth nodes are assigned to hub 2, and the third node is assigned to hub 4, is represented with the matrix:

$$X = \begin{pmatrix} 0 & 1 & 0 & 0 & 0 \\ 0 & 1 & 0 & 0 & 0 \\ 0 & 0 & 0 & 1 & 0 \\ 0 & 0 & 0 & 1 & 0 \\ 0 & 1 & 0 & 0 & 0 \end{pmatrix} \quad \text{is represented by the following array} \quad s = \begin{pmatrix} 2 & 4 & | 1 & 2 & 1 \end{pmatrix}$$

This array has the following meaning: As p is equal 2, the first two positions within s, represent the hubs (2 and 4). Then, these two values are eliminated from L and $L = \{1,3,5\}$. The last three positions on the array s are the assignments of the remaining nodes in L. That is, nodes 1 and 5 are assigned to hub 2 and the node 3 is assigned to hub 4. Note that this representation may produce feasible and infeasible solutions for the capacitated problem.

The solutions for the capacitated problem should take into account the distribution capacity of each node. This capacity does not be exceeded when a hub is established. On this way, when a new solution is formed or there are movements, the capacity of the hubs should be considered to support the distributed flow. New variables are introduced to study this fact:

Total flow array: saves the total flow with origin and destiny at node i, and is defined

$$T_i = \sum_{j \in N} W_{ij} + \sum_{j \in N} W_{ji}$$

Capacity vector: Saves the capacity of each node and it is symbolized by V.

These two arrays are sufficient to control the feasibility of the problem.

On our work, the solutions feasibility is studied at each step of the process to built a new solution. That is, each time that a new node i is going to be assigned to a *hub j*, and nodes k, l and m have been previously assigned to j, the procedure considers if *hub j* admits the total flow of node i. Afterwards, the equation $T_k+T_l+T_m+T_i < V_j$ must be satisfied. In other case the node i could not be assigned to *hub j* and it should be assigned to another *hub*. If i can be assigned to none of the *hubs*, the current solution will be a non feasible solution.

4.2 Population Creation Method

The initial population of solutions is generated using a Multistart method. The method operates as follows:

1. Generate a random solution and improve it with a greedy procedure in the location phase and a greedy procedure in the allocation phase. The greedy procedure in location was used previously in [9] with the name of LS1, and it is based on the resolution of the 1-hub problems and 1-exchange moves. Only new solutions are introduced in P.

2   If the improved solution does not belong to P, P is updated with this new solution.
3   The steps 1 and 2 are repeated until the elements in P are P_size.

4.3 Reference Set Generation Method

RS is initialized with the best b_1 solutions of P. Then, the b_2 solutions in $P - RS$ that are the most disparate (dispersed) with respect to RS are added to RS. The solution t is defined to be the most disparate solution with respect to the solutions already in RS, if t maximizes the distance function:

$$d(t, RS) = \min\{d(t, s)/s \in RS\}$$

with:

$$d(t,s) = \rho \cdot d\_loc(t, s) + (1 - \rho) \cdot d\_as(t, s)$$

where:

$d\_loc(t, s) =$ number of different hubs in s and t
$d\_as(t, s) =$ number of nodes with different assignment in both solutions

4.4 Subset Generation Method

We consider all the subsets of 2 solutions of the reference set. This is the stopping rule of the pseudocode (figure 1).

4.5 Relinking Method

The relinking between each pair of solutions into each subset of the reference set is achieved with two neighbourhood structures defined below. Moves within these neighbourhoods determine the location relinking and leave the allocation relinking to a random strategy. The underlying idea is that if we try to preserve the common information in location, the allocation relinking phase will be dominated by the location relinking.

Location Relinking Neighbourhood: This neighbourhood provides moves to build new solutions by choosing (at random) a position from the range $[1 \ldots p - 1]$ (where p is the length of the location array), and preserving the places that are on the left of this position. The moves of the neighbourhood are structured so that they yield feasible outcomes at each step. In particular, if we consider an instance of the problem with $n = 15$ and $p = 5$, and if we take the solutions with location arrays $l_1 = (7\ 9\ |\ 10\ 11\ 15)$, and $l_2 = (1\ 3\ |\ 8\ 9\ 10)$, where | symbolizes the interchange point, then the new solution is produced as follows:

- marks the common hubs of l_1 and l_2:
    $l_1 = (1\ 3\ |\ 8\ \underline{9\ 10})$ $l_2 = (7\ \underline{9}\ |\ \underline{10}\ 11\ 15)$
- places on the first positions (ordered) the non marked symbols:
    $l_1 = (1\ 3\ |\ 8\ \underline{9\ 10})$ $l_2 = (7\ 11\ |\ 15\ \underline{9\ 10})$
- switchs the two tail portions:
    $l_1 = (1\ 3\ |\ 15\ \underline{9\ 10})$ $l_2 = (7\ 11\ |\ 8\ \underline{9\ 10})$
- removes the marks and reorder the resulting arrays:
    $l_1 = (1\ 3\ 9\ 10\ 15)$ $l_2 = (7\ 8\ 9\ 10\ 11)$.

Allocation Relinking Neighbourhood: This neighbourhood is defined by reference to moves of the following type. For each pair of allocation arrays, generate a random position in the range $[1 \dots n - p - 1]$, and simultaneously execute a move from s_1 towards s_2 and from s_2 towards s_1 by swapping the two tail portions. For example, if we consider the instance used in the location relinking and if the two allocation–arrays to be operated are $a_1 = (1\ 1\ 2\ 3\ 1\ 5\ 2\ 3\ 5\ 4)$ and $a_2 = (2\ 1\ 3\ 1\ 4\ 2\ 5\ 5\ 2\ 3)$, the new solutions are performed with the following process:

- generate a random position in both arrays:
$$a_1 = (1\ 1\ 2\ 3\ 1\ 5\ |\ 2\ 3\ 5\ 4) \qquad a_2 = (2\ 1\ 3\ 1\ 4\ 2\ |\ 5\ 5\ 2\ 3)$$
- switch the two tail portions:
$$a_1 = (1\ 1\ 2\ 3\ 1\ 5\ |\ 5\ 5\ 2\ 3) \qquad a_2 = (2\ 1\ 3\ 1\ 4\ 2\ |\ 2\ 3\ 5\ 4)$$
- then, the new allocation arrays are:
$$a_1 = (1\ 1\ 2\ 3\ 1\ 5\ 5\ 5\ 2\ 3) \qquad a_2 = (2\ 1\ 3\ 1\ 4\ 2\ 2\ 3\ 5\ 4).$$

4.6 Improvement Method

As an improvement method, we use a greedy procedure both for location and alloca-tion. In [9], the greedy procedure proved to be very effective. The improvement method is applied only to the solutions with the large incoming and outcoming flow.

4.7 Reference Set Update Method

At the end of each iteration, the RS is updated with the best solution (according to the objective value) or the most diversified solution (according to the distance function defined above).

5   GRASP

GRASP (Greedy Randomized Adaptive Search Procedure) [3] is a constructive meta-heuristic consisting of two phases. In the *constructive phase*, a solution is iteratively constructed randomly selecting one element of the restricted candidate list. Then, in the *post-processing phase*, it is attempted to improve this solution using a improved method (generally, a descent local search). These steps are repeated until a stopping rule is satisfied. The best overall solution is kept as the result.

5.1 *GRASP* for the Capacitated Single Allocation *p – hub* Median Problem

5.1.1 Greedy Randomized Function

The myopic greedy location function should build the solution iteratively, adding one element to it at each time. In the case of the capacitated single allocation *p-hub* me-dian problem, the greedy evaluation function is divided in two phases. A location phase and an allocation phase.

In the location phase one element is added at each time to the set of hubs. As many myopic criteria might be proposed, we have adapted the results obtained by Klince-wicz [8] for the *p-hub* location problem, using the true improvement S_i in the objec-tive function that would result if a node i were added to the set of nodes.

If we denote $\tilde{C}_j$ as the minimum cost of assigning j to a node in the current set of hubs, the value of S_i is calculated as follows:

$$S_i = \sum_{j \in N} \left(O_j + D_j \right) \cdot \left(\tilde{C}_j - C_{ji} \right)$$

At each step of the construction phase, the nodes i with largest value of S_i are kept in a candidate list. Then, a random selection is performed among the elements of the candidate list and the selected node is added to the set of hubs.
The allocation phase consists of an allocation to the nearest hub.

5.1.2 Local Search
As the Local Search, we use a greedy procedure for location and an allocation to the nearest hub. The greedy procedure used in this case has been defined in 4.2. and was previously formulated in [9].

6 *Path Relinking* with *GRASP* for the Capacitated Single Allocation *p-hub* Median Problem

The combination of *GRASP* and *PR* consists of the use of *GRASP* as the Population Creation Method of the *PR*. On this manner, we start the *PR* with a set of elite solutions to combine them.

7 Computational Experience

The computational experience compares the behaviour of the hybrid *GRASP & PR* method for the capacitated *p-hub* median problem, against the *PR* procedure. The algorithms have been implemented using the C language programming and they have been executed on a 2400 Mhz Intel Pentium IV PC, with 1 Gb of RAM, running the Linux Operating System.

Our tests are performed over the AP Problem set (see [2]), corresponding to the Australian postal office between 200 major cities. To the best of our knowledge, these problems have not been solved in the literature. So, our tables present the comparison between *PR* and *GRASP+PR* and the computational results obtained for problems with $n = 10, 20, 25, 40$ and 50, $p = 2, 3, 4$ and 5, and with $n = 100$, $p = 5, 10, 15, 20$.

Since both procedures are random heuristics, they were executed 30 times on each problem. The problem we consider on this paper is study from the computacional point of view. This kind of study has not been proposed previously on the literature. So, the optimal value of the objective function is not known for each parameters combination.

For each problem, the tables show the best solution obtained, and the results for both heuristics (time measured in minutes and seconds, and the factor ϕ – in percent – by which the objective function exceeds the best solution obtained during the execution):

$$\phi = 100 \cdot \left(\frac{objective - best\ solution}{best\ solution} \right)$$

Table 1. Computational Results for the AP problem

			PR				GRASP + PR			
				ϕ / CPU time				ϕ / CPU time		
n	P	Best cost obtained	%	Min	μ	Max	%	Min	μ	Max
10	2	35476.08	100	0.00	0.00	0.00	100	0.00	0.00	0.00
				0.11	0.18	0.25		0.15	0.22	0.39
	3	28635.25	100	0.00	0.00	0.00	100	0.00	0.00	0.00
				0.17	0.20	0.23		0.15	0.20	0.26
	4	30793.10	100	0.00	0.00	0.00	100	0.00	0.00	0.00
				0.18	0.19	0.22		0.18	0.21	0.27
	5	30080.613	100	0.00	0.00	0.00	100	0.00	0.00	0.00
				0.21	0.26	0.28		0.22	0.27	0.31
20	2	29885.14	100	0.00	0.00	0.00	100	0.00	0.00	0.00
				0.25	0.30	0.32		0.28	0.35	0.39
	3	29752.69	97	0.06	0.09	0.11	100	0.00	0.03	0.07
				0.23	0.29	0.34		0.25	0.35	0.38
	4	28550.62	100	0.00	0.00	0.00	100	0.00	0.00	0.00
				0.25	0.29	0.33		0.29	0.32	0.38
	5	24858.77	100	0.00	0.00	0.00	100	0.00	0.00	0.00
				0.27	0.36	0.41		0.28	0.35	0.40
25	2	37359.92	100	0.00	0.00	0.00	100	0.00	0.00	0.00
				1.33	1.39	1.42		0.35	0.38	0.42
	3	42811.63	100	0.00	0.00	0.00	100	0.00	0.00	0.00
				1.33	2.39	2.41		0.35	0.40	0.43
	4	24900.25	90	0.05	0.07	0.15	100	0.00	0.00	0.00
				1.39	2.46	3.49		0.39	0.45	0.47
	5	24169.54	97	0.02	0.08	0.11	97	0.00	0.06	0.08
				1.36	3.40	4.48		0.37	0.41	0.45
40	2	46734.69	100	0.00	0.00	0.00	100	0.00	0.00	0.00
				2.38	2.46	3.49		0.38	0.45	0.47
	3	59999.84	100	0.00	0.00	0.00	100	0.00	0.00	0.00
				2.44	3.48	5.51		0.41	0.45	0.48
	4	51628.19	90	0.03	0.06	0.11	100	0.00	0.00	0.00
				2.48	5.50	6.55		0.45	0.46	0.49
	5	47761.54	90	0.03	0.08	0.11	90	0.00	0.01	0.03
				3.49	5.52	8.55		0.44	0.48	0.51
50	2	52655.26	80	0.03	0.08	0.11	100	0.00	0.03	0.09
				4.55	5.45	10.5		0.52	0.55	0.58
	3	51445.13	90	0.06	0.09	0.15	97	0.00	0.03	0.08
				5.05	5.58	11.3		0.51	0.53	0.56
	4	49673.34	80	0.08	0.13	0.16	90	0.00	0.04	0.06
				5.28	6.01	10.8		0.49	0.55	1.01
	5	49245.18	70	0.09	0.11	0.18	97	0.00	0.05	0.08
				18.3	20.1	29.3		1.35	1.52	2.15
100	5	85698.27	70	0.06	0.09	0.15	90	0.00	0.02	0.05
				15.3	16.2	25.0		1.15	1.18	1.25
	10	82735.11	50	0.02	0.04	0.09	80	0.00	0.04	0.07
				16.5	18.2	28.3		1.11	1.14	1.22
	15	75699.27	50	0.08	0.11	0.19	80	0.00	0.04	0.09
				25.7	28.3	33.4		2.18	12.2	18.3
	20	80003.89	30	0.11	0.23	0.27	70	0.00	0.03	0.08
				29.3	35.3	38.3		2.27	12.8	23.2

In the table below, the column labelled *PR* presents the results obtained for the *PR* procedure and the columns labelled *GRASP&PR* give the results of the hybrid *GRASP&PR* method. For both heuristics minimum *(Min)*, average *(μ)* and maximum *(Max)* values on the 30 executions are showed. The column labelled %, represents the number of times (on percentage), that the heuristic reached the best solution obtained. On the *PR*, the size of the population was $P = 10$, and the $|RS| = b = 10$. These sizes were tuned experimentally.

The number of iterations of the *GRASP* algorithm was 10 when it was executed alone and 5 when it was executed in combination with *PR*. The candidate list used in *GRASP* is formed by 10 solutions. At the same time, the number of iterations of the *PR* procedure was fixed between 5 and 10, both when the heuristic was executed alone and when it was combined on the hybrid *GRASP & PR*. The AP problems were chosen to yield a test set containing large problem instances.

As the table below show, the hybrid *GRASP & PR* procedure provides, in general terms, a better behaviour than the unique *PR*. Although the running times are very similar and, sometimes they are better for *PR* than for the hybrid algorithm, the quality of the solutions are better for the hybrid *GRASP & PR* for most of the cases, this is not true only in one case. The variance of *GRASP & PR* is also lower than the *PR* version. This low variance gives the method interesting robustness properties what allows to reduce the number of executions. This reduction of the variance is mainly due to the high quality solutions provided by the *GRASP* as initial population.

In conclusion, the hybrid heuristic *GRASP & PR* developed for the capacitated *p-hub* median problem performs better than *PR* overall, and performs significantly better than *PR* on large problem instances. We anticipate the possibility of further improvement of our procedure by additional refinement of the neighbourhoods used to define moves for combining solutions. In particular, our current neighbourhoods are designed for executing a single large step move in each, to transform the parents into offspring for the next iteration. Neighbourhoods that transition between the solutions using additional intermediate steps are also appropriate to consider.

Acknowledgments

This paper has been partially supported by the Spanish Ministry of Science and Technology through the projects TIC2002-04242-C03-01, TIC2002-04400-C03-03; 70% of which are FEDER founds and, for the Canary Government project PI042004/088.

References

1. Andreas T. Ernst and Mohan Krishnamoorthy, Efficient Algorithms for the Uncapacitated Single Allocation *p-hub* Median Problem. Location Science, vol. 4, No.3, (1995), 130 – 154.
2. Beasley, J.E. "OR-library" http://mscmga.ms.ic.ac.uk/info.html.
3. T.A. Feo and M.G.C. Resende. Greedy randomized adaptive search procedures. Journal of Global Optimization, 6 (1995) 109-133.
4. Glover, F. Heuristics for Integer Programming Using Surrogate Constraints. Decision Sciences, Vol 8, No 1, (1977), 156 – 166.
5. Glover, F. Tabu Search for Nonlinear and Parametric Optimization (with Links to Genetic Algorithms), Discrete Applied Mathematics, 49 (1994), 231-255.
6. Glover, F., M. Laguna and R. Marti. Fundamentals of Scatter Search and *Path Relinking*. Control and Cybernetics, Vol. 29, No. 3 (2000), 653-684.
7. Klincewicz, J.G. Heuristics for the *p-hub* location problem. European Journal of Operational Research, Vol. 53 (1991), 25 – 37.

8. Klincewicz, J.G. Avoiding local optima in the *p-hub* location problem using tabu search and *GRASP*. Annals of Operation Research, Vol. 40 (1992), 283 – 302.
9. M. Pérez, F. Almeida, J. M. Moreno – Vega. Fast Heuristics for the *p-hub* Location Problem, presented in EWGLAX, Murcia (Spain), (1998).
10. M. Pérez, F. Almeida, J. M. Moreno – Vega. Genetic Algorithm with Multistart Search for the *p-Hub* Median Problem. Proceedings of EUROMICRO 98. IEEE Computer Society (2000), 702-707.
11. O'Kelly, M. (1986). The location of interacting hub facilities, Transportation Science, Vol 20, (1986), 92 – 106.
12. Skorin-Kapov, D & Skorin-Kapov, J. (1994). "On tabu search for the location of interacting hub facilities". European Journal of Operations Research, Vol. 73, 502-509.
13. Ernst, A.T., H. Hamacher, H. Jiang, M. Krishnamoorthy and G. Woeginger (2002). Uncapacitated Single and Multiple Allocation *p-Hub* Center Problems", to appear.
14. Ernst, A.T., M. Krishnamoorthy (1999). Solution algorithms for the capacitated single allocation hub location problem. Annals of Operations Research, Vol. 86, 141-159.
15. Turgut Ayking (1994). Lagrangian relaxation based approaches to capacitated hub and spoke network design problem. European Journal of Operational Research, Vol. 77, 1-23.
16. Manuel Laguna, Rafael Martí (2003). Scatter Search. Kluwer Academic Publishers.

Author Index

Lecture Notes in Computer Science

For information about Vols. 1–3534

please contact your bookseller or Springer